CONTENTS.

Introduction Page 1

CHAPTER I.

Feeble resources of civilized man in a desert—Ross Cox, Peter the Wild Boy, and the Savage of Aveyron—A Moskito Indian on Juan Fernandez—Conditions necessary for the production of utility 6

CHAPTER II.

Society a system of exchanges—Security of individual property the principle of exchange—Alexander Selkirk and Robinson Crusoe—Imperfect appropriation and unprofitable labour 14

CHAPTER III.

Adventures of John Tanner—Habits of the American Indians—Their sufferings from famine, and from the absence among them of the principle of division of labour—Evils of irregular labour—Respect to property—Their present improved condition—Hudson's Bay Indians 23

CHAPTER IV.

The Prodigal—Advantages of the poorest man in civilized life over the richest savage—Savings-banks, deposits, and interest—Progress of accumulation—Insecurity of capital, its causes and results—Property, its constituents—Accumulation of capital 38

CHAPTER V.

Common interests of Capital and Labour—Labour directed by Accumulation—Capital enhanced by Labour—Balance of rights and duties—Relation of demand and supply—Money exchanges—Intrinsic and representative value of money 49

CHAPTER VI.

Importance of capital to the profitable employment of labour—Contrast between the prodigal and the prudent man: the Dukes of Buckingham and Bridgewater—Making good for trade—Unprofitable consumption—War against capital in the middle ages—Evils of corporate privileges—Condition of the people under Henry VIII. 60

CHAPTER VII.

Rights of labour—Effects of slavery on production—Condition of the Anglo Saxons—Progress of freedom in England—Laws regulating labour—Wages and prices—Poor-law—Law of settlement 71

CHAPTER VIII.

Possessions of the different classes in England—Condition of Colchester in 1301—Tools, stock-in-trade, furniture, &c.—Supply of food—Comparative duration of human life—Want of facilities for commerce—Plenty and civilization not productive of effeminacy—Colchester in the present day 82

CHAPTER IX.

Certainty the stimulus to industry—Effects of insecurity—Instances of unprofitable labour—Former notions of commerce—National and class prejudices, and their remedy 96

CHAPTER X.

Employment of machinery in manufactures and agriculture—Erroneous notions formerly prevalent on this subject—Its advantages to the labourer—Spade-husbandry—The principle of machinery—Machines and tools—Change in the condition of England consequent on the introduction of machinery—Modern New Zealanders and ancient Greeks—Hand-mills and water-mills 106

CHAPTER XI.

Present and former condition of the country—Progress of cultivation—Evil influence of feudalism—State of agriculture in the sixteenth century—Modern improvements—Prices of wheat—Increased breadth of land under cultivation—Average consumption of wheat—Implements of agriculture now in use—Number of agriculturists in Great Britain 124

CHAPTER XII.

Production of a knife—Manufacture of iron—Raising coal—The hot-blast—Iron bridges—Rolling bar-iron—Making steel—Sheffield manufactures—Mining in Great Britain—Numbers engaged in mines and metal manufactures 139

CHAPTER XIII.

Conveyance and extended use of coal—Consumption at various periods—Condition of the roads in the seventeenth and eighteenth centuries—Advantages of good roads—Want of roads in Australia—Turnpike-roads—Canals—Railway of 1680—Railway statistics 157

CHAPTER XIV.

Houses—The Pyramids—Mechanical power—Carpenters' tools—American machinery for building—

Bricks—Slate—Household fittings and furniture—Paper-hangings—Carpets—Glass—Pottery—Improvements effected through the reduction or repeal of duties on domestic requirements 174

CHAPTER XV.

Dwellings of the people—Oberlin—The Highlander's candlesticks—Supply of water—London waterworks—Street-lights—Sewers 199

CHAPTER I.

Feeble resources of civilized man in a desert—Ross Cox, Peter the Wild Boy, and the Savage of Aveyron—A Moskito Indian on Juan Fernandez—Conditions necessary for the production of utility.

Let us suppose a man brought up in civilized life, cast upon a desert land—without food, without clothes, without fire, without tools. We see the human being in the very lowest state of helplessness. Most of the knowledge he had acquired would be worse than useless; for it would not be applicable in any way to his new position. Let the land upon which he is thrown produce spontaneous fruits—let it be free from ferocious animals—let the climate be most genial—still the man would be exceedingly powerless and wretched. The first condition of his lot, to enable him to maintain existence at all, would be that he should labour. He must labour to gather the berries from the trees—he must labour to obtain water from the rivulets—he must labour to form a garment of leaves, or of some equally accessible material, to shield his body from the sun—he must labour to render some cave or hollow tree a secure place of shelter from the dews of night. There would be no intermission of the labour necessary to provide a supply of food from hand to mouth, even in the season when wild fruits were abundant. If this labour, in the most favourable season, were interrupted for a single day, or at most for two or three days, by sickness, he would in all probability perish. But, when the autumn was past, and the wild fruits were gone, he must prolong existence as some savage tribes are reported to do—by raw fish and undressed roots. The labour of procuring these would be infinitely greater than that of climbing trees for fruit. To catch fish without nets, and scratch up roots with naked hands, is indeed painful toil. The helplessness of this man's condition would principally be the effect of one circumstance;—he would possess no accumulation of former labour by which his present labour might be profitably directed. *The power of labour would in his case be in its least productive state.* He would partly justify the assertion that man has the feeblest natural means of any animal;—because he would be utterly unpossessed of those means which the reason of man has accumulated around every individual in the social state.

We asked the reader to *suppose* a civilized man in the very lowest state in which the power of labour may be exercised, because there is no record of any civilized man being for any length of time in such a state.

Ross Cox, a Hudson's Bay trader, whose adventures were given to the world some twenty years ago, was in this state for a fortnight; and his sufferings may furnish some idea of the greater miseries of a continuance in such a powerless condition. Having fallen asleep in the woods of the north-west of America, which he had been traversing with a large party, he missed the traces of his companions. The weather being very hot, he had left nearly all his clothes with his horse when he rambled from his friends. He had nothing to defend himself against the wolves and serpents but a stick; he had nothing of which to make his bed but long grass and rushes; he had nothing to eat but hips and wild cherries. The man would doubtless have perished, unless he had met with some Indians, who knew better how to avail themselves of the spontaneous productions around them. But this is not an instance of the continuance of Labour in the lowest state of its power.

The few individuals, also, who have been found exposed in forests, such as Peter the Wild Boy, and the Savage of Aveyron,—who were discovered, the one about a century ago, in Germany, the other about forty years since, in France,—differed from the civilized man cast naked upon a desert shore in this particular—their *wants* were of the lowest nature. They were not raised above the desires of the most brutish animals. They supplied those desires after the fashion of brutes. Peter was enticed from the woods by the sight of two apples, which the man who found him displayed. He did not like bread, but he eagerly peeled green sticks, and chewed the rind. He had, doubtless, subsisted in this way in the woods. He would not, at first, wear shoes, and delighted to throw the hat which was given him into the river. He was brought to England, and lived many years with a farmer in Hertfordshire. During the Scotch Rebellion, in 1745, he wandered into Norfolk; and having been apprehended as a suspicious character, was sent to prison. The gaol was on fire; and Peter was found in a corner, enjoying the warmth of the flames without any fear. The Savage of Aveyron, in the same manner, had the lowest desires and the feeblest powers. He could use his hands, for instance, for no other purpose than to seize upon an object; and his sense of touch was so defective, that he could not distinguish a raised surface, such as a carving, from a painting. This circumstance of the low physical and intellectual powers of these unfortunate persons prevents us exhibiting them as examples of the state which we asked the reader to suppose.

Let us advance another step in our view of the power of Labour. Let us take a man in one respect in the same condition that we supposed—left upon a desert land, without any direct social aid; but with some help to his labour by a small Accumulation of former industry. We have instances on record of this next state.

In the year 1681 a Moskito Indian was left by accident on the island of Juan Fernandez, in the Pacific Ocean; the English ship in which he was a sailor having been chased off the coast by some hostile Spanish vessels. Captain Dampier describes this man's condition in the following words:—

> "This Indian lived here alone above three years; and although he was several times sought after by the Spaniards, who knew he was left on the island, yet they could never find him. He was in the woods hunting for goats, when Captain Watlin drew off his men, and the ship was under sail before he came back to shore. He had with him his gun, and a knife, with a small horn of powder, and a few shot; which being spent, he contrived a way, by notching his knife, to saw the barrel of his gun into small pieces, wherewith he made harpoons, lances, hooks, and a long knife; heating the pieces first in the fire, which he struck with his gun-flint, and a piece of the barrel of his gun, which he hardened, having learnt to do that among the English. The hot pieces of iron he would hammer out and bend as he pleased with stones, and saw them with his jagged knife, or grind them to an edge by long labour, and harden them to a good temper as there was occasion.[5] With such instruments as he made in that manner, he got such provisions as the island afforded, either goats or fish. He told us that at first he was forced to eat seal, which is very ordinary meat, before he had made hooks; but afterwards he never killed any seals but to make lines, cutting their skins into thongs. He had a little house, or hut, half a mile from the sea, which was lined with goat's skin; his couch, or barbecu of sticks, lying along about two feet distance from the ground, was spread with the same, and was all his bedding. He had no clothes left, having worn out those he brought from Watlin's ship, but only a skin about his waist. He saw our ship the day before we came to an anchor, and did believe we were English; and therefore killed three goats in the morning, before we came to an anchor, and dressed them with cabbage, to treat us when we came ashore."

Here, indeed, is a material alteration in the wealth of a man left on an uninhabited island. He had a regular supply of goats and fish; he had the means of cooking this food; he had a house lined with goats' skins, and bedding of the same; his body was clothed with skins; he had provisions in abundance to offer, properly cooked, when his old companions came to him after three years' absence. What gave him this power to labour profitably? —to maintain existence in tolerable comfort? Simply, the gun, the knife, and the flint, which he accidentally had with him when the ship sailed away.

The flint and the bit of steel which he hardened out of the gun-barrel gave him the means of procuring fire; the gun became the material for making harpoons, lances, and hooks, with which he could obtain fish and flesh. Till he had these tools, he was compelled to eat seal's flesh. The instant he possessed the tools, he could make a selection of what was most agreeable to his taste. It is almost impossible to imagine a human being with less accumulation about him. His small stock of powder and shot was soon spent, and he had only an iron gun-barrel and a knife left, with the means of changing the form of the gun-barrel by fire. Yet this single accumulation enabled him to direct his labour, as all labour is directed even in its highest employment, to the change of form and change of place of the natural supplies by which he was surrounded. He created nothing; he only gave his natural supplies a value by his labour. Until he laboured, the things about him had no value, as far as he was concerned; when he did obtain them by labour, they instantly acquired a value. He brought the wild goat from the mountain to his hut in the valley—he changed its place; he converted its flesh into cooked food, and its skin into a lining for his bed—he changed its form. Change of form and change of place are the beginning and end of all human labour; and the Moskito Indian only employed the same principle for the supply of his wants which directs the labour of all the producers of civilized life into the channels of manufactures or commerce.

But the Moskito Indian, far removed as his situation was above the condition of the man without any accumulation of former labour—that is, of the man without any capital about him—was only *in the second stage in which the power of labour can be exercised*, and in which it is comparatively still weak and powerless. He laboured—he laboured with accumulation—but he laboured without that other power which gives the last and highest direction to profitable labour.

Let us state all the conditions necessary for the production of Utility, or of what is essential to the support, comfort, and pleasure of human life:—

1. *That there shall be Labour*.

The man thrown upon a desert island without accumulation,—the half-idiot boy who wandered into the German forests at so early an age that he forgot all the usages of mankind,—were each compelled to labour, and to labour unceasingly, to maintain existence. Even with an unbounded command of the spontaneous productions of nature, this condition is absolute. It applies to the inferior animals as well as to man. The bee wanders from flower to flower, but it is to labour for the honey. The sloth hangs upon the branches of a tree, but he labours till he has devoured all the leaves, and then climbs another tree. The condition of the support of animation is labour; and if the labour of all animals were miraculously suspended for a season, very short as compared with the duration of individual life, the reign of animated nature upon this globe would be at an end.

African Hut.

The second condition in the production of utility is,—

2. *That there shall be accumulation of former labour, or Capital.*

Without accumulation, as we have seen, the condition of man is the lowest in the scale of animal existence. The reason is obvious. Man requires some accumulation to aid his natural powers of labouring; for he is not provided with instruments of labour to anything like the perfection in which they exist amongst the inferior animals. He wants the gnawing teeth, the tearing claws, the sharp bills, the solid mandibles that enable quadrupeds, and birds, and insects to secure their food, and to provide shelter in so many ingenious ways, each leading us to admire and reverence the directing Providence which presides over such manifold contrivances. He must, therefore, to work profitably, accumulate instruments of work. But he must do more, even in the unsocial state, where he is at perfect liberty to direct his industry as he pleases, uncontrolled by the rights of other men. He must accumulate stores of covering and of shelter. He must have a hut and a bed of skins, which are all accumulations, or capital. He must, further, have a stock of food, which stock, being the most essential for human wants, is called *provisions*, or things provided. He would require this provision against the accidents which may occur to his own health, and the obstacles of weather, which may prevent him from fishing or hunting. The lowest savages have some stores. Many of the inferior animals display an equal care to provide for the exigencies of the future. But still, all such labour is extremely limited. When a man is occupied only in providing immediately for his own wants—doing everything for himself, consuming nothing but what he produces himself,—his labour must have a very narrow range. The supply of the lowest necessities of our nature can only be attended to, and these must be very ill supplied. The Moskito Indian had fish, and goats' flesh, and a rude hut, and a girdle of skins; and his power of obtaining this wealth was insured to him by the absence of other individuals who would have been his competitors for what the island spontaneously produced. Had other Indians landed in numbers on the island, and had each set about procuring everything for himself, as the active Moskito did, they would have soon approached the point of starvation; and then each would have begun to plunder from the other, unless they had found out the principle that would have given them all plenty. There wanted, then, another power to give the labour of the Indian a profitable direction, besides that of accumulation. It is a power which can only exist where man is social, as it is his nature to be;—and where the principles of civilization are in a certain degree developed. It is, indeed, the beginning and the end of all civilization. It is itself civilization, partial or complete. It is the last and the most important condition in the production of useful commodities,—

3. *That there shall be exchanges.*

There can be no exchanges without accumulation—there can be no accumulation without labour. Exchange is that step beyond the constant labour and the partial accumulation of the lower animals, which makes man the lord of the world.

CHAPTER II.

Society a system of exchanges—Security of individual property the principle of exchange—Alexander Selkirk and Robinson Crusoe—Imperfect appropriation and unprofitable labour.

Society, both in its rudest form and in its most refined and complicated relations, is nothing but a system of Exchanges. An exchange is a transaction in which both the parties who make the exchange are benefited;—and, consequently, society is a state presenting an uninterrupted succession of advantages for all its members. Every time that we make a free exchange we have a greater desire for the thing which we receive than for the thing which we give;—and the person with whom we make the exchange has a greater desire for that which we offer him than for that which he offers us. When one gives his labour for wages, it is because he has a higher estimation of the wages than of the profitless ease and freedom of remaining unemployed;—and, on the contrary, the employer who purchases his labour feels that he shall be more benefited by the results of that labour than by retaining the capital which he exchanges for it. In a simple state of society, when one man exchanges a measure of wheat for the measure of wine which another man possesses, it is evident that the one has got a greater store of wheat than he desires to consume himself, and that the other, in the same way, has got a greater store of wine;—the one exchanges something to eat for something to drink, and the other something to drink for something to eat. In a refined state of society, when money represents the value of the exchanges, the exchange between the abundance beyond the wants of the possessor of one commodity and of another is just as real as the barter of wheat for wine. The only difference is, that the exchange is not so direct, although it is incomparably more rapid. But, however the system of exchange be carried on,—whether the value of the things exchanged be determined by barter or by a price in money,—all the exchangers are benefited, because all obtain what they want, through the store which they possess of what they do not want.

It has been well said that "Man might be defined to be an animal that makes exchanges."[6] There are other animals, indeed, such as bees and ants amongst insects, and beavers amongst quadrupeds, which to a certain extent are social; that is, they concur together in the execution of a common work for a common good: but as to their individual possessions, each labours to obtain what it desires from sources accessible to all, or plunders the stores of others. Not one insect or quadruped, however wonderful may be its approaches to rationality, has the least idea of making a formal exchange with another. The modes by which the inferior animals communicate their thoughts are probably not sufficiently determinate to allow of any such agreement. The very foundation of that agreement is a complicated principle, which man alone can understand. It is the Security of individual Property. Immediately that this principle is established, labour begins to work profitably, for it works with exchange. If the principle of appropriation were not acted upon at all, there could be no exchange, and consequently no production. The scanty bounty of nature might be scrambled for by a few miserable individuals—and the strongest would obtain the best share; but this insecurity would necessarily destroy all accumulation. Each would of course live from hand to mouth, when the means of living were constantly exposed to the violence of the more powerful. This is the state of the lowest savages, and as it is an extreme state it is a rare one,—no security, no exchange, no capital, no labour, no production. Let us apply the principle to an individual case.

The poet who has attempted to describe the feelings of a man suddenly cut off from human society, in "Verses supposed to be written by Alexander Selkirk during his solitary abode in the island of Juan Fernandez," represents him as saying, "I am monarch of all I survey."[7] Alexander Selkirk was left upon the same island as the Moskito Indian; and his adventures there have formed the groundwork of the beautiful romance of "Robinson Crusoe." The meaning of the poet is, that the unsocial man had the same right over all the natural productive powers of the country in which he had taken up his abode, as we each have over light and air. He was alone; and therefore he exercised an absolute although a barren sovereignty, over the wild animals by which he was surrounded—over the land and over the water. He was, in truth, the one proprietor—the one capitalist, and the one labourer—of the whole island. His absolute property in the soil, and his perfect freedom of action, were both dependent upon one condition—that he should remain alone. If the Moskito Indian, for instance, had remained in the island, Selkirk's entire sovereignty must have been instantly at an end. Some more definite principle of appropriation must have been established, which would have given to Selkirk, as well as to the Moskito Indian, the right to appropriate distinct parts of the island each to his particular use. Selkirk, for example, might have agreed to remain on the eastern coast, while the Indian might have established himself on the western; and then the fruits, the goats, and the fish of the eastern part would have been appropriated to Selkirk, as distinctly as the clothes, the musket, the iron pot, the can, the hatchet, the knife, the mathematical instruments, and the Bible which he brought on shore.[8] If the Indian's territory had produced something which Selkirk had not, and if Selkirk's land had also something which the Indian's had not, they might have become exchangers. They would have passed into that condition naturally enough;—imperfectly perhaps, but still as easily as any barbarous people who do not cultivate the earth, but exchange her spontaneous products.

The poet goes on to make the solitary man say, "My right there is none to dispute." The condition of Alexander Selkirk was unquestionably one of absolute liberty. His rights were not measured by his duties. He had all rights and no duties. Many writers on the origin of society have held that man, upon entering into union with his fellow-men, and submitting, as a necessary consequence of this union, to the restraints of law and government, sacrifices a portion of his liberty, or natural power, for the security of that power which remains to him. No such agreement amongst mankind could ever have possibly taken place; for man is by his nature, and without any agreement, a social being. He is a being whose rights are balanced by the uncontrollable force of their relation to the rights of others. The succour which the infant man requires from its parents, to an extent, and for a duration, so much exceeding that required for the nurture of other creatures, is the natural beginning of the social state, established insensibly and by degrees. The liberty which the social man is thus compelled by the force of circumstances to renounce amounts only to a restraint upon his brute power of doing injury to his fellow-men: and for this sacrifice, in itself the cause of the highest individual and therefore general good, he obtains that dominion over every other being, and that control over the productive forces of nature, which alone can render him the monarch of all he surveys. The poor sailor, who for four years was cut off from human aid, and left alone to struggle for the means of supporting existence, was an exception, and a very rare one, to the condition of our species all over the world. His absolute rights placed him in the condition of uncontrolled feebleness; if he had become social, he would have put on the regulated strength of rights balanced by duties.

Alexander Selkirk was originally left upon the uninhabited island of Juan Fernandez at his own urgent desire. He was unhappy on board his ship, in consequence of disputes with his captain; and he resolved to rush into a state which might probably have separated him for ever from the rest of mankind. In the belief that he should be so separated, he devoted all his labour and all his ingenuity to the satisfaction of his own wants alone. By continual exercise, he was enabled to run down the wild goat upon the mountains; and by persevering search, he knew where to find the native roots that would render his goat's flesh palatable. He never thought, however, of providing any store beyond the supply of his own personal necessities. He had no motive for that thought; because there was no human being within his reach with whom he might exchange that store for other stores. The very instant, however, that the English ships, which finally gave him back to society, touched upon his shores,—before he communicated by speech with any of his fellow-men, or was discovered by them,—he became social. He saw that he must be an exchanger. Before the boat's crew landed he had killed several goats, and prepared a meal for his expected guests. He knew that he possessed a commodity which they did not possess. He had fresh meat, whilst they had only salt. Of course what he had to offer was acceptable to the sailors; and he received in exchange protection, and a place amongst them. He renounced his sovereignty, and became once more a subject. It was better for him, he thought, to be surrounded with the regulated power of civilization, than to wield at his own will the uncertain strength of solitary uncivilization. But, had he chosen to remain upon his island, as in after-years he regretted he had not done, although a solitary man he would not have been altogether cut off from the hopes and the duties of the social state. If he had chosen to remain after that visit from his fellow-men, he would have said to them, before they had left him once more alone, "I have hunted for you my goats, I have dug for you my roots, I have shown you the fountains which issue out of my rocks;—these are the resources of my dominion: give me in exchange for them a fresh supply of gunpowder and shot, some of your clothes, some of the means of repairing these clothes, some of your tools and implements of cookery, and more of your books to divert my solitary hours." Having enjoyed the benefits which he had bestowed, they would, as just men, have paid the debt which they had incurred, and the exchange would have been completed. Immediately that they had quitted his shores, Selkirk would have looked at his resources with a new eye. His hut was rudely fashioned and wretchedly furnished. He had fashioned, and furnished it as well as he could by his own labour, working upon his own materials. The visit which he had received from his fellow-men, after he had abandoned every hope of again looking upon their faces, would have led him to think that other ships would come, with whose crews he might make other exchanges,—new clothes, new tools, new materials, received as the price of his own accumulations. To make the best of his circumstances when that day should arrive, he must redouble his efforts to increase his stock of commodities,—some for himself, and some to exchange for other commodities, if the opportunity for exchange should ever come. He must therefore transplant his vegetables, so as to be within instant reach when they should be wanted. He must render his goats domestic, instead of chasing them upon the hills. He must go forward from the hunting state, into the pastoral and agricultural.

Robinson Crusoe. (From a design by Stothard.)

In Defoe's story, Robinson Crusoe is represented as going into this pastoral and agricultural state. But he had more resources than Selkirk; and he at last obtained one resource which carried him back, however incompletely, into the social condition. He acquired a fellow-labourer. He made a boat by his own unassisted labour; but he could not launch it. When Friday came, and was henceforth his faithful friend and willing servant, he could launch his boat. Crusoe ultimately left his island; for the boat had given him a greater command over his circumstances. But had he continued there in companionship with Friday, there must have been such a compact as would have prevented either struggling for the property which had been created. The course of improvement that we have imagined for Selkirk supposes that he should continue in his state of exclusive proprietor—that there should be none to dispute his right. If other ships had come to his shores—if they had trafficked with him from time to time—exchanged clothes and household conveniences, and implements of cultivation, for his goats' flesh and roots—it is probable that other sailors would in time have desired to partake his plenty;—that a colony would have been founded—that the island would have become populous. It is perfectly clear that, whether for exchange amongst themselves, or for exchange with others, the members of this colony could not have stirred a step in the cultivation of the land without appropriating its produce;—and they could not have appropriated its produce without appropriating the land itself. Cultivation of the land for a common stock would have gone to the establishment precisely of the same principle;—they would still have been exchangers amongst themselves, and the partnership would not have lasted a day, unless each man's share of what the partnership produced had been rendered perfectly secure to him. Without security they could not have accumulated—without accumulations they could not have exchanged—without exchanges they could not have carried forward their labours with any compensating productiveness.

Imperfect appropriation—that is, an appropriation which respects personal wealth, such as the tools and conveniences of an individual, and even secures to him the fruits of the earth when he has gathered them, but which has not reached the last step of a division of land—imperfect appropriation such as this raises up the same invincible obstacles to the production of utility; because, with this original defect, there must necessarily be unprofitable labour, small accumulation, limited exchange. Let us exemplify this by another individual case.

We have seen, in the instances of the Moskito Indian and of Selkirk, how little a solitary man can do for himself, although he may have the most unbounded command of natural supplies—although not an atom of those natural supplies, whether produced by the earth or the water, is appropriated by others—when, in fact, he is monarch of all he surveys. Let us trace the course of another man, advanced in the ability to subdue all things to his use by association with his fellow-men; but carrying on that association in the rude and unproductive relations of savage life;—not desiring to "replenish the earth" by cultivation, but seeking only to appropriate the means of existence which it has spontaneously produced;—labouring, indeed, and exchanging, but not labouring and exchanging in a way that will permit the accumulation of wealth, and therefore remaining poor and miserable. We are not about to draw any fanciful picture, but merely to select some facts from a real narrative.

[6] Dr. Whately's Lectures on Political Economy.

[7] Cowper's Miscellaneous Poems.

[8] These circumstances are recorded in Captain Woodes Rogers' Cruising Voyage round the World, 1712.

CHAPTER III.

Adventures of John Tanner—Habits of the American Indians—Their sufferings from famine, and from the absence among them of the principle of division of labour—Evils of irregular labour—Respect to property—Their present improved condition—Hudson's Bay Indians.

In the year 1828 there came to New York a white man named John Tanner, who had been thirty years a captive amongst the Indians in the interior of North America. He was carried off by a band of these people when he was a little boy, from a settlement on the Ohio river, which was occupied by his father, who was a clergyman. The boy was brought up in all the rude habits of the Indians, and became inured to the abiding miseries and uncertain pleasures of their wandering life. He grew in time to be a most skilful huntsman, and carried on large dealings with the agents of the Hudson's Bay Company, in the skins of beavers and other animals which he and his associates had shot or entrapped. The history of this man was altogether so curious, that he was induced to furnish the materials for a complete narrative of his adventures; and, accordingly, a book, fully descriptive of them, was prepared for the press by Dr. Edwin James, and printed at New York, in 1830. It is of course not within the intent of our little work to furnish any regular abridgment of John Tanner's story; but it is our wish to direct attention to some few particulars, which appear to us strikingly to illustrate some of the positions which we desire to enforce, by thus exhibiting their practical operation.

The country in which this man lived so many years is that immense territory belonging to the United States, which at that period was covered by boundless forests which the progress of civilization had not then cleared away. In this region a number of scattered Indian tribes maintained a precarious existence by hunting the moose-deer and the buffalo for their supply of food, and by entrapping the foxes and martens of the woods and the beavers of the lakes, whose skins they generally exchanged with the white traders of Europe for articles of urgent necessity, such as ammunition and guns, traps, axes, and woollen blankets; but too often for ardent spirits, equally the curse of savage and of civilized life. The contact of savage man with the outskirts of civilization perhaps afflicts him with the vices of both states. But the principle of exchange, imperfectly and irregularly as it operated amongst the Indians, furnished some excitement to their ingenuity and their industry. Habits of providence were thus to a certain degree created; it became necessary to accumulate some capital of the commodities which could be rendered valuable by their own labour, to exchange for commodities which their own labour, without exchange, was utterly unable to procure. The principle of exchange, too, being recognised amongst them in their dealings with foreigners, the security of property—without which, as we have shown, that principle cannot exist at all—was one of the great rules of life amongst themselves. But still these poor Indians, from the mode which they proposed to themselves for the attainment of property, which consisted only in securing what nature had produced, without directing the course of her productions, were very far removed from the regular attainment of those blessings which civilized society alone offers. We shall exemplify these statements by a few details.

Dying Lion

The country over which these people ranged occupies a surface that may be roughly described as five or six times as large as all England. They had the unbounded command of all the natural resources of that country; and yet their entire numbers did not equal the population of a moderately sized English county. It may be fairly said that each Indian required a thousand acres for his maintenance. The supplies of food were so scanty—a scantiness which would at once have ceased to exist had there been any cultivation—that if a large number of these Indians assembled together to co-operate in their hunting expeditions, they were very soon dispersed by the urgent desire of satisfying hunger. Tanner says, "We all went to hunt beavers in concert. In hunts of this kind the proceeds are sometimes equally divided; but in this instance every man retained what he had killed. In three days I collected as many skins as I could carry. But in these distant and hasty hunts little meat could be brought in; and the whole band was soon suffering with hunger. Many of the hunters, and I among others, for want of food became extremely weak, and unable to hunt far from home." What an approach is this to the case of the lower animals; and how forcibly it reminds us of the passage in Job (c. iv., v. 11), "The fierce lion perisheth for lack of prey."[9] In another place he says, "I began to be dissatisfied at remaining with large bands of Indians, as was usual for them, after having remained a short time in a place, to suffer from hunger." These sufferings were not, in many cases, of short duration, or of trifling intensity. Tanner describes one instance of famine in the following words:—"The Indians gathered around, one after another, until we became a considerable band, and then we began to suffer from hunger. The weather was very severe, and our suffering increased. A young woman was the first to die of hunger. Soon after this, a young man, her brother, was taken with that kind of delirium or madness which precedes death in such as die of starvation. In this condition he had left the lodge of his debilitated and desponding parents; and when, at a late hour in the evening, I returned from my hunt, they could not tell what had become of him. I left the camp about the middle of the night, and, following his track, I found him at some distance, lying dead in the snow."

This worst species of suffering equally existed at particular periods, whether food was sought for by large or by small parties, by bands or by individuals. Tanner was travelling with the family of the woman who had adopted him. He says, "We had now a short season of plenty; but soon became hungry again. It often happened that for two or three days we had nothing to eat; then a rabbit or two, or a bird, would afford us a prospect of protracting the sufferings of hunger a few days longer." Again he says, "Having subsisted for some time almost entirely on the inner bark of trees, and particularly of a climbing vine found there, our strength was much reduced."

The misery which is thus so strikingly described proceeded from the circumstance that the labour of the Indians did not take a profitable direction; and that this waste of labour (for unprofitable applications of labour are the greatest of all wastes) arose from the one fact, that in certain particulars these Indians laboured without appropriation. They depended upon the chance productions of nature, without compelling her to produce; and they did not compel her to produce, because there was no appropriation of the soil, the most efficient natural instrument of production. If the Indians had directed the productive powers of the earth to the growth of corn, instead of to the growth of foxes' skins, they would have become rich. But they could not have reached this point without appropriation of the soil. They had learnt the necessity of appropriating the products of the soil, when they had bestowed labour upon obtaining them; but the last step towards productiveness was not taken. The Indians therefore were poor; the European settlers who had taken this last step were rich.

The imperfect appropriation which existed amongst the Indians, preventing, as it did, the accumulation of capital, prevented the application of that skill and knowledge which is preserved and accumulated by the Division of employment. Tanner describes a poor fellow who was wounded in the arm by the accidental discharge of a gun. As there was little surgical skill amongst the community, because no one could devote himself to the business of surgery, the Indian, as the only chance of saving his life, resolved to cut off his own arm; "and taking two knives, the edge of one of which he had hacked into a sort of saw, he with his right hand and arm cut off his left, and threw it from him as far as he could." The labour which an individual must go through when the state of society is so rude that there is scarcely any division of employment, and consequently scarcely any exchanges, is exhibited in many passages of Tanner's narrative. We select one. "I had no pukkavi, or mats for a lodge, and therefore had to build one of poles and long grass. I dressed more skins, made my own mocassins and leggings, and those for my children; cut wood and cooked for myself and family, made my snow-shoes, &c. &c. All the attention and labour I had to bestow about home sometimes kept me from hunting, and I was occasionally distressed for want of provisions. I busied myself about my lodge in the night-time. When it was sufficiently light I would bring wood, and attend to other things without; at other times I was repairing my snow-shoes, or my own or my children's clothes. For nearly all the winter I slept but a very small part of the night."

Tanner was thus obliged to do everything for himself, and consequently to work at very great disadvantage, because the principle of exchange was so imperfectly acted upon by the people amongst whom he lived. This principle of exchange was imperfectly acted upon, because the principle of appropriation was imperfectly acted upon. The occupation of all, and of each, was to hunt game, to prepare skins, to sell them to the traders, to make sugar from the juice of maple-trees, to build huts, and to sew the skins which they dressed and the blankets which they bought into rude coverings for their bodies. Every one of them did all of these things for himself, and of course he did them very imperfectly. The people were not divided into hunters, and furriers, and dealers, and sugar-makers, and builders, and tailors. Every man was his own hunter, furrier, dealer, sugar-maker, builder, and tailor; and consequently, every man, like Tanner, was so occupied by many things, that want of food and want of rest were ordinary sufferings. He describes a man who was so borne down and oppressed by these manifold wants, that, in utter despair of being able to surmount them, he would lie still till he was at the point of starvation, replying to those who tried to rouse him to kill game, that he was too poor and sick to set about it. By describing himself as poor, he meant to say that he was destitute of all the necessaries and comforts whose possession would encourage him to add to the store. He had little capital. The skill which he possessed of hunting game gave him a certain command over the spontaneous productions of the forest; but, as his power of hunting depended upon chance supplies of game, his labour necessarily took so irregular a direction, and was therefore so unproductive, that he never accumulated sufficient for his support in times of sickness, or for his comfortable support at any time. He became, therefore, despairing; and had that perfect apathy, that indifference to the future, which is the most pitiable evidence of extreme wretchedness. This man felt his powerless situation more keenly than his companions; but with all savage tribes there is a want of steady and persevering exertion, proceeding from the same cause. Severe labour is succeeded by long fits of idleness, because their labour takes a chance direction. This is a universal case. Habits of idleness, of irregularity, of ferocity, are the characteristics of all those who maintain existence by the pursuit of the unappropriated productions of nature; while constant application, orderly arrangement of time, and civility to others, result from systematic industry. The savage and the poacher are equally the slaves of violent impulses—equally disgusted at the prospect of patient application. When the support of life depends upon chance supplies, the reckless spirit of a gambler is sure to take possession of the whole man; and the misery which results from these chance supplies produces either dejection or ferocity. The author of this book used to observe the habits of a class of such persons, who frequent the Thames at Eton; and he thus described them in verses of his boyhood:—

> What boat is this which creeps so lazily
> Up the still stream? How quietly falls the drip
> Of the slow paddle! Now it shoots along,
> As if that lone man fear'd us. Well I ken
> His rough and dangerous trade. He knows each hole
> Where the quick-sighted eel delights to swim
> When clouds obscure the moon; and there he lays
> His traps and gins, and then he sleeps awhile;
> But rouses up before the prying dawn
> Betrays his course; and out he cautiously glides
> To try his doubtful luck. Perchance he finds
> Stores that may buy him bread; but oft'ner still
> His toil is fruitless, and deject he comes
> Home to his emberless hearth, and sits him down,
> Idle and starving through the busy day.

Mungo Park describes the wretched condition of the inhabitants of countries in Africa where small particles of gold are found in the rivers. Their lives were spent in hunting for the gold to exchange for useful commodities, instead of raising the commodities themselves; and they were consequently poor and miserable, listless and unsteady. Their fitful industry had too much of chance mixed up with it to afford a certain and general profit. The accounts which of late years we have received from the gold-diggings of California and Australia exhibit the same suffering from the same cause. The natives of Cape de la Hogue, in Normandy, were the most wretched and ferocious people in all France, because they depended principally for support on the wrecks that were frequent on their coasts. When there were no tempests, they made an easy transition from the character of wreckers to that of robbers. A benefactor of his species taught these unhappy people to collect a marine plant to make potash. They immediately became profitable labourers and exchangers; they obtained a property in the general intelligence of civilized life; the capital of society raised them from misery to wealth, from being destroyers to being producers.

The Indians, as we thus see, were poor and wretched, because they had no appropriation beyond articles of domestic use; because they had no property in land, and consequently no cultivation. Yet even they were not insensible to the importance of the principle, for the preservation of the few advantages that belonged to their course of life. Tanner says, "I have often known a hunter leave his traps for many days in the woods, without visiting them, or feeling any anxiety about their safety." The Indians even carried the principle of appropriation almost to a division of land; for each tribe, and sometimes each individual, had an allotted hunting-ground—imperfectly appropriated, indeed, by the first comer, and often contested with violence by other hunters, but still showing that they approached the limit which divides the savage from the civilized state, and that, if cultivation were introduced amongst them, there would be a division of land, as a matter of necessity. The security of individual property is the foundation of all social improvement. It is impossible to speak of the productive power of labour in the civilized state, without viewing it in connexion with that great principle of society which considers all capital as appropriated.

When 'Capital and Labour' was written twenty years ago, the Indian tribes who were abiding in the territory of the United States were principally in the condition which has been described by Tanner. The want of resources in the country of the Indians was so manifest, that, when commissioners from the government of the United States, in 1802, gathered together the chiefs of the various tribes of the Creek Indians in their own country, to propose to them a plan for their civilization, it became necessary to provide for the support of the people so assembled by conveying food into the forests from the stores of the American towns. The Indians have now vanished from their old hunting-grounds. Where they so recently maintained a precarious existence, there are populous cities, navigable rivers, roads, railways. The clink of the hammer is heard in the forge, and the rush of the stream from the mill-dam tells of agriculture and commerce. But even the Indians themselves have become labourers. They have been removed to a large tract of country, far away from the settled parts of the United States, and have been raised into the dignity of cultivators. The Cherokees, the Creeks, and the Choctaws, with many smaller tribes, now breed cattle instead of hunting martens. They have houses in the place of huts; they have schools and churches. Instead of being extirpated by famine or the sword, they have been adopted into the great family of civilized man.

Penn's Treaty with the Indians.

But this wise and humane arrangement of the United States has not wholly removed the Indians from the wide regions of North America. In the Hudson's Bay territories the life which Tanner described still goes forward. The wants of civilized society—the desire to possess the earth—have transported the Indians from the banks of the Ohio to the lands watered by the Arkansas. The opposite principle has retained them on the shores of Hudson's Bay. They are wanted there as hunters, and are not encouraged as cultivators. They are kept out of the pale of civilization, and not received within it. The rude industry of the Hudson's Bay Indians is stimulated by the luxury of Europe into an employ which would cease to exist if the people became civilized. If agriculture were introduced amongst them—if they were to grow corn and keep domestic animals—they would cease to be hunters of foxes and martens, because their wants would be much better supplied by other modes of labour, involving less suffering and less uncertainty. As it is, the traders, who want skins, do not think of giving the Indians tools to work the ground, and seeds to put in it, and cows and sheep to breed other cows and sheep. They avail themselves of the uncivilized state of these poor tribes, to render them the principal agents in the manufacture of fur, to supply the luxuries of another hemisphere. But still the exchange which the hunters carry on with the European traders, imperfect as it is in all cases, and unjust as it is in many, is better for the Indians than no exchange; although we fear that ardent spirits take away from the Indians the greater number of the advantages which would otherwise remain with them as exchangers. If the Indians had no skins to give to Europe, Europe would have no blankets and ammunition to give to them. They would obtain their food and clothing by the use of the bow alone. They would live entirely from hand to mouth. They would have no motive for accumulation, because there would be no exchanges; and they would consequently be even poorer and more helpless than they are now as exchangers of skins. They are suffering from the effects of small accumulations and imperfect exchange; but these are far better than no accumulation and no exchange. If the course of their industry were to be changed by perfect appropriation—if they were consequently to become cultivators and manufacturers, instead of wanderers in the woods to hunt for wild and noxious animals—they would, in the course of years, have abundance of profitable labour, because they would have abundance of capital. This is the better lot of the tribes with whom the government of the

United States has made a far nobler treaty than Penn made with his Indians. As it is, their accumulations are so small, that they cannot proceed with their own uncertain labour of hunting without an advance of capital on the part of the traders; and thus, even [Pg 33] [Pg 34] in the rude tradings of these poor Indians, credit, that complicated instrument of commercial exchange, operates upon the direction of their labour. Of course credit would not exist at all without appropriation. Their rights of property are perfect as far as they go; but they are not carried far enough to direct their labour into channels which would ensure sufficient production for the labourers. Their labour is unproductive because they have small accumulations;—their accumulations are small because they have imperfect exchange;—their exchange is imperfect because they have limited appropriation. We may illustrate this state of imperfect production by another passage from Tanner's story:—

> "The Hudson's Bay Company had now no post in that part of the country, and the Indians were soon made conscious of the advantage which had formerly resulted to them from the competition between rival trading companies. Mr. Wells, at the commencement of winter, called us all together, gave the Indians a ten-gallon keg of rum and some tobacco, telling them at the same time he would not credit one of them the value of a single needle. When they brought skins he would buy them, and give in exchange such articles as were necessary for their comfort and subsistence during the winter. I was not with the Indians when this talk was held. When it was reported to me, and a share of the presents offered me, I not only refused to accept anything, but reproached the Indians for their pusillanimity in submitting to such terms. They had been accustomed for many years to receive credits in the fall; they were now entirely destitute not of clothing merely, but of ammunition, and many of them of guns and traps. How were they, without the accustomed aid from the traders, to subsist themselves and their families during the ensuing winter? A few days afterwards I went to Mr. Wells, and told him that I was poor, with a large family to support by my own exertions; and that I must unavoidably suffer, and perhaps perish, unless he would give me such a credit as I had always in the fall been accustomed to receive. He would not listen to my representation, and told me roughly to be gone from his house. I then took eight silver beavers, such as are worn by the women as ornaments on their dress, and which I had purchased the year before at just twice the price that was commonly given for a capote;[10] I laid them before him on the table, and asked him to give me a capote for them, or retain them as a pledge for the payment of the price of the garment, as soon as I could procure the peltries.[11] He took up the ornaments, threw them in my face, and told me never to come inside of his house again. The cold weather of the winter had not yet set in, and I went immediately to my hunting-ground, killed a number of moose, and set my wife to make the skins into such garments as were best adapted to the winter season, and which I now saw we should be compelled to substitute for the blankets and woollen clothes we had been accustomed to receive from the traders."

This incident at once shows us that the great blessing of the civilized state is its increase of the powers of production. Here we see the Indians, surrounded on all sides by wild animals whose skins might be made into garments, reduced to the extremity of distress because the traders refused to advance them blankets and other necessaries, to be used during the months when they were employed in catching the animals from which they might obtain the skins. It is easy to see that the Indians were a long way removed from the power of making blankets themselves. Before they could reach this point, their forests must have been converted into pasture-grounds;— they must have raised flocks of sheep, and learnt all the various complicated arts, and possessed all the ingenious machinery, for converting wool into cloth. By their exchange of furs for blankets, they obtained a share in the productiveness of civilization;—they obtained comfortable clothing with much less labour than they could have made it out of the furs. If Tanner had not considered the capote which he desired to obtain from the traders, better, and less costly, than the garment of moose-skins, he would not have carried on any exchange of the two articles with the traders. The skins of martens and foxes were only valuable to the Indians, without exchange, for the purpose of sewing together to make covering. They had a different value in Europe as articles of luxury; and therefore the Indians by exchange obtained a greater plenty of superior clothing than if they had used the skins themselves. But the very nature of the trade, depending upon chance supplies, rendered it impossible that they should accumulate. They had such pressing need of ammunition, traps, and blankets, that the produce of the labour of one hunting season was not more than sufficient to procure the commodities which they required to consume in the same season. But supposing the Indians could have bred foxes and martens and beavers, as we breed rabbits, for the supply of the European demand for fur, doubtless they would have then advanced many steps in the character of producers. The thing is perhaps impossible; but were it possible, and were the Indians to have practised it, they would immediately have become capitalists, to an extent that would have soon rendered them independent of the credit of the traders. They must, however, have previously established a more perfect appropriation. Each must have enclosed his own hunting-ground; and each must have raised some food for the maintenance of his own stock of beavers, foxes, and martens. It would be easier, doubtless, to raise the food for themselves, and ultimately to exchange corn for clothing, instead of furs for clothing. When this happens—and it will happen sooner or later, unless the remnant of the hunting Indians are extirpated by their poverty, which proceeds from their imperfect production—Europe must go without the brilliant variety of skins which we procure at the cost of so much labour, accompanied with so much wretchedness, because the labour is so unproductive to the labourers. When the ladies of London and Paris are compelled to wear boas of rabbits instead of sables, and when the hair of the beaver ceases to be employed in the manufacture of our hats, the wooded regions of Hudson's Bay will have been cleared—the fur-bearing animals will have perished—corn will be growing in the forest and the marsh—the inhabitants will be building houses instead of trapping foxes;—there will be appropriation and capital, profitable labour and comfort. Three hundred thousand mink and marten-skins will no longer be sent from those shores to London in one year; but Liverpool may send to those shores woven cottons and worsteds, pottery and tools, in exchange for products whose cultivation will have exterminated the minks and martens.

CHAPTER IV.

The Prodigal—Advantages of the poorest man in civilized life over the richest savage—Savings-banks, deposits, and interest—Progress of accumulation—Insecurity of capital, its causes and results—Property, its constituents—Accumulation of capital.

There is an account in Foster's Essays of a man who, having by a short career of boundless extravagance dissipated every shilling of a large estate which he inherited from his fathers, obtained possession again of the whole property by a course which the writer well describes as a singular illustration of decision of character. The unfortunate prodigal, driven forth from the home of his early years by his own imprudence, and reduced to absolute want, wandered about for some time in a state of almost unconscious despair, meditating self-destruction, till he at last sat down upon a hill which overlooked the fertile fields that he once called his own. "He remained," says the narrative, "fixed in thought a number of hours, at the end of which he sprang from the ground with a vehement exulting emotion. He had formed his resolution, which was, that all these estates should be his again; he had formed his plan, too, which he instantly began to execute." We shall show, by and by, how this plan worked in detail;—it will be sufficient, just now, to examine the principles upon which it was founded. He looked to no freak of fortune to throw into his lap by chance what he had cast from him by wilfulness. He neither trusted to inherit those lands from their present possessor by his favour, nor to wring them from him by a course of law. He was not rash and foolish enough to dream of obtaining again by force those possessions which he had exchanged for vain superfluities. But he resolved to become once more their master by the operation of the only principle which could give them to him in a civilized society. He resolved to obtain them again by the same agency through which he had lost them—by exchange. But what had he to exchange? His capital was gone, even to the uttermost farthing; he must labour to obtain new capital. With a courage worthy of imitation he resolved to accept the very first work that should be offered to him, and, however low the wages of that work, to spend only a part of those wages, leaving something for a store. The day that he made this resolution he carried it into execution. He found some service to be performed—irksome, doubtless, and in many eyes degrading. But he had a purpose which made every occupation appear honourable, as every occupation truly is that is productive of utility. Incessant labour and scrupulous parsimony soon accumulated for him a capital; and the store, gathered together with such energy, was a rapidly increasing one. In no very great number of years the once destitute labourer was again a rich proprietor. He had earned again all that he had lost. The lands of his fathers were again his. He had accomplished his plan.

A man so circumstanced—one who possesses no capital, and is only master of his own natural powers—if suddenly thrown down from a condition of ease, must look upon the world, at the first view, with deep apprehension. He sees everything around him appropriated. He is in the very opposite condition of Alexander Selkirk, when he is made to exclaim "I am monarch of all I survey." Instead of feeling that his "right there is none to dispute," he knows that every blade of corn that covers the fields, every animal that grazes in the pastures, is equally numbered as the property of some individual owner, and can only pass into his possession by exchange. In the towns it is the same as in the country. The dealer in bread and in clothes,— the victualler from whom he would ask a cup of beer and a night's lodging, —will give him nothing, although they will exchange everything. He cannot exist, except as a beggar, unless he puts himself in the condition to become an exchanger.

But still, with all these apparent difficulties, his prospects of subsisting, and of subsisting comfortably, are far greater than in any other situation in which he must labour to live. As we have already seen, the condition of by far the greater number of the millions that constitute the exchangers of civilized society is greatly superior to that of the few thousands who exist upon the precarious supplies of the unappropriated productions of nature in the savage life. Although an exchange must always be made—although in very few cases "the fowl and the brute" offer themselves to the wayfaring man for his daily food—although no herbs worth the gathering can be found for the support of life in the few uncultivated parts of a highly cultivated country—the aggregate riches are so abundant, and the facilities which exist for exchanging capital for labour are therefore so manifold, that the poorest man in a state of civilization has a much greater certainty of supplying all his wants, and of supplying them with considerably more ease, than the richest man in a state of uncivilization. The principle upon which he has to rely is, that in a highly civilized country there is large production. There is large production because there is profitable labour;—there is profitable labour because there is large accumulation;—there is large accumulation because there is unlimited exchange;—there is unlimited exchange because there is universal appropriation. John Tanner was accounted a rich man by the Indians—doubtless because he was more industrious than the greater number of them; but we have seen what privations he often suffered. He suffered privations because there was no capital, no accumulation of the products of labour, in the country in which he lived. Where such a store exists, the poorest man has a tolerable certainty that he may obtain his share of it as an exchanger; and the greater the store the greater the certainty that his labour, or power of adding to the store, will obtain a full proportion of what previous labour has gathered together.

In 1853 the amount of stock vested to the account of depositors in savings-banks in the United Kingdom was 34,546,434*l*. Since the establishment of savings-banks, 68,885,283*l*. had been so invested; and the gross amount of interest paid to the depositors had been 25,733,771*l*. This large capital, which had so fructified as to produce more than twenty-five millions as interest, was an accumulation, penny by penny, shilling by shilling, and pound by pound, of the savings of that class of persons who, in every country, have the greatest difficulty in accumulating. Habitual efforts of self-denial, and a rigid determination to postpone temporary gratification to permanent good, could alone have enabled these accumulators to retain so much of what they had produced beyond the amount of what they consumed.

The capital sum of more than thirty-four millions now belonging to the depositors in 575 savings-banks, represents as many products of industry as could be bought by that sum. It is a capital which remains for the encouragement of *productive* consumption; that is, it is now applied as a fund for setting others to produce,—to enable them to consume while they produce,—and in like manner to accumulate some part of their productions beyond what they consume. The millions of interest which the depositors have received is the price paid for the use of the capital by others who require its employment. The whole amount of our national riches—the capital of this and of every other country—has been formed by the same slow but certain process of individual savings, and the accumulations of savings, stimulating new industry, and yielding new accumulations.

The consumption of any production is the destruction of its value. The production was created by industry to administer to individual wants, to be consumed, to be destroyed. When a thing capable of being consumed is produced, a value is created; when it is consumed, that value is destroyed. The general mass of riches then remains the same as it was before that production took place. If the power to produce, and the disposition to consume, were equal and constant, there could be no saving, no accumulation, no capital. If mankind, by their intelligence, their skill, their division of employments, their union of forces, had not put themselves in a condition to produce more than is consumed while the great body of industrious undertakings is in progress, society would have been stationary, —civilization could never have advanced.

It may assist us in making the value of capital more clear, if we take a rapid view of the most obvious features of the accumulation of a highly civilized country.

The first operation in a newly settled country is what is termed to clear it. Look at a civilized country, such as England. It *is* cleared. The encumbering woods are cut down, the unhealthy marshes are drained. The noxious animals which were once the principal inhabitants of the land are exterminated; and their place is supplied with useful creatures, bred, nourished, and domesticated by human art, and multiplied to an extent exactly proportioned to the wants of the population. Forests remain for the produce of timber, but they are confined within the limits of their utility;—mountains "where the nibbling flocks do stray," have ceased to be barriers between nations and districts. Every vegetable that the diligence of man has been able to transplant from the most distant regions is raised for food. The fields are producing a provision for the coming year; while the stock for immediate consumption is ample, and the laws of demand and supply are so perfectly in action, and the facilities of communication with every region so unimpeded, that scarcity seldom occurs, and famine never. Rivers have been narrowed to bounds which limit their inundations, and they have been made navigable wherever their navigation could be profitable. The country is covered with roads, with canals, and now, more especially, with railroads, which render distant provinces as near to each other for commercial purposes as neighbouring villages in less advanced countries. Science has created the electric telegraph, by which prices are equalized through every district, by an instant communication between producers and consumers. Houses, all possessing some comforts which were unknown even to the rich a few centuries ago, cover the land, in scattered farm-houses and mansions, in villages, in towns, in cities, in capitals. These houses are filled with an almost inconceivable number of conveniences and luxuries—furniture, glass, porcelain, plate, linen, clothes, books, pictures. In the stores of the merchants and traders the resources of human ingenuity are displayed in every variety of substances and forms that can exhibit the multitude of civilized wants; and in the manufactories are seen the wonderful adaptations of science for satisfying those wants at the cheapest cost. The people who inhabit such a civilized land have not only the readiest communication with each other by the means of roads and canals, but can trade by the agency of ships with all parts of the world. To carry on their intercourse amongst themselves they speak one common language, reduced to certain rules, and not broken into an embarrassing variety of unintelligible dialects. Their written communications are convoyed to the obscurest corners of their own country, and even to the most remote lands, with prompt and unfailing regularity, and now with a cheapness which makes the poorest and the richest equal in their power to connect the distant with their thoughts by mutual correspondence. Whatever is transacted in such a populous hive, the knowledge of which can afford profit or amusement to the community, is recorded with a rapidity which is not more astonishing than the general accuracy of the record. What is more important, the discoveries of science, the elegancies of literature, and all that can advance the general intelligence, are preserved and diffused with the utmost ease, expedition, and security, so that the public stock of knowledge is constantly increasing. Lastly, the general well-being of all is sustained by laws—sometimes indeed imperfectly devised and expensively administered, but on the whole of infinite value to every member of the community; and the property of all is defended from external invasion and from internal anarchy by the power of government, which will be respected only in proportion as it advances the general good of the humblest of its subjects, by securing their capital from plunder and defending their industry from oppression.

This capital is ready to be won by the power of every man capable of work. But he must exercise this power in complete subjection to the natural laws by which every exchange of society is regulated. These laws sometimes prevent labour being instantly exchanged with capital, for an exchange necessarily requires a balance to be preserved between what one man has to supply and what another man has to demand; but in their general effect they secure to labour the certainty that there shall be abundance of capital to exchange with; and that, if prudence and diligence go together, the labourer may himself become a capitalist, and even pass out of the condition of a labourer into that of a proprietor, or one who lives upon accumulated produce. The experience of every day shows this process going forward—not in a solitary instance, as in that of the ruined and restored man whose tale we have just told, but in the case of thriving tradesmen all around us, who were once servants. But if the labourer or the great body of labourers were to imagine that they could obtain such a proportion of the capital of a civilized country except as exchangers, the store would instantly vanish. They might perhaps divide by force the crops in barns and the clothes in warehouses—but there would be no more crops or clothes. The principle upon which all accumulation depends, that of security of property, being destroyed, the accumulation would be destroyed. Whatever tends to make the state of society insecure, tends to prevent the employment of capital. In despotic countries, that insecurity is produced by the tyranny of one. In other countries, where the people, having been misgoverned, are badly educated, that insecurity is produced by the tyranny of many. In either case, the bulk of the people themselves are the first to suffer, whether by the outrages of a tyrant, or by their own outrages. They prevent labour, by driving away to other channels the funds which support labour. In some eastern countries, where, when a man becomes rich, his property is seized upon by the one tyrant, nobody dares to avow that he has any property. Capital is not employed; it is hidden: and the people who have capital live not upon its profits, but by the diminution of the capital itself. In the very earliest times we hear of concealed riches. In the book of Job those who "long for death" are said to "dig for it more than for hid treasures." The tales of the East are full of allusions to money buried and money dug up. The poor woodman, in making up his miserable faggot, discovers a trap-door, and becomes rich. In India, where the rule of Mohammedan princes was usually one of tyranny, even now the search after treasure goes on. The popular mind is filled with the old traditions; and so men dream of bags of gold to be discovered in caves and places of desolation, and they forthwith dig, till hope is banished, and the real treasure is found in systematic industry. It was the same in the feudal times in England, when the lord tyrannized over his vassals, and no property was safe but in the hands of the strongest. In those times people who had treasure buried it. Who thinks of burying treasure now in England? In the plays and story-books which depict the manners of our own early times, we constantly read of people finding bags of money. We never find bags of money now, except when a very old hoard, hidden in some time of national trouble, comes to light. So little time ago as the reign of Charles II. we read of the Secretary to the Admiralty going down from London to his country-house, with all his money in his carriage, to bury it in his garden. What Samuel Pepys records of his doings with his own money, was a natural consequence of the practices of a previous time. He also chronicles, in several places of his curious Diary, his laborious searches, day by day, for 7000*l.* hid in butter firkins in the cellars of the Tower of London. Why is money not hidden and not sought for now? Because people have security for the employment of it, and by the employment of it in creating new produce the nation's stock of capital goes on hourly increasing. When we read in Blackstone's 'Commentaries on the Laws of England,' that the concealment of treasure-trove, or found treasure, from the king, is a misdemeanour punishable by fine and imprisonment, and that it was formerly a capital offence, we at once see that this is a law no longer for our time; and we learn from this instance, as from many others, how the progress of civilization silently repeals laws which belong to another condition of the people.

Treasure-finding.

When we look at the nature of the accumulated wealth of society, it is easy to see that the poorest member of it who dedicates himself to profitable labour is in a certain sense rich—rich, as compared with the unproductive and therefore poor individuals of any uncivilized tribe. The very scaffolding, if we may so express it, of the social structure, and the moral forces by which that structure was reared, and is upheld, are to him riches. To be rich is to possess the means of supplying our wants—to be poor is to be destitute of those means. Riches do not consist only of money and lands, of stores of food or clothing, of machines and tools. The particular knowledge of any art—the general understanding of the laws of nature—the habit from experience of doing any work in the readiest way—the facility of communicating ideas by written language—the enjoyment of institutions conceived in the spirit of social improvement—the use of the general conveniences of civilized life, such as roads—these advantages, which the poorest man in England possesses or may possess, constitute individual property. They are means for the supply of wants, which in themselves are essentially more valuable for obtaining his full share of what is appropriated, than if all the productive powers of nature were unappropriated, and if, consequently, these great elements of civilization did not exist. Society obtains its almost unlimited command over riches by the increase and preservation of knowledge, and by the division of employments, including union of power. In his double capacity of a consumer and a producer, the humblest man has the full benefit of these means of wealth—of these great instruments by which the productive power of labour is carried to its highest point.

But if these common advantages, these public means of society, offering so many important agents to the individual for the gratification of his wants, alone are worth more to him than all the precarious power of the savage state—how incomparably greater are his advantages when we consider the wonderful accumulations, in the form of private wealth, which are ready to be exchanged with the labour of all those who are in a condition to add to the store. It has been truly said by M. Say, a French economist, "It is a great misfortune to be poor, but it is a much greater misfortune for the poor man to be surrounded only with other poor like himself." The reason is obvious. The productive power of labour can be carried but a very little way without accumulation of capital. In a highly civilized country, capital is heaped up on every side by ages of toil and perseverance. A succession, during a long series of years, of small advantages to individuals unceasingly renewed and carried forward by the principle of exchanges, has produced this prodigious amount of the aggregate capital of a country whose civilization is of ancient date. This accumulation of the means of existence, and of all that makes existence comfortable, is principally resulting from the labours of those who have gone before us. It is a stock which was beyond their own immediate wants, and which was not extinguished with their lives. It is our capital. It has been produced by labour alone, physical and mental. It can be kept up only by the same power which has created it, carried to the highest point of productiveness by the arrangements of society.

CHAPTER V.

Common interests of Capital and Labour—Labour directed by Accumulation—Capital enhanced by Labour—Balance of rights and duties—Relation of demand and supply—Money exchanges—Intrinsic and representative value of money.

There is an old proverb, that "When two men ride on one horse, one man must ride behind." Capital and Labour are, as we think, destined to perform a journey together to the end of time. We have shown how they proceed on this journey. We have shown that, although Labour is the parent of all wealth, its struggles for the conversion of the rude supplies of nature into objects of utility are most feeble in their effects till they are assisted by accumulation. Before the joint interests of Labour and Capital were at all understood, they kept separate; when they only began to be understood, as we shall show, they were constantly pulling different ways, instead of giving "a long pull, a strong pull, and a pull altogether;" and even now, when these interests in many respects are still imperfectly understood, they occasionally quarrel about the conditions upon which they will continue to travel in company. In the very outset of the journey, Labour, doubtless, took the lead. In the dim morning of society Labour was up and stirring before Capital was awake. Labour did not then ride; he travelled slowly on foot through very dirty ways. Capital, at length, as slowly followed after, through the same mire, but at an humble distance from his parent. But when Capital grew into strength, he saw that there were quicker and more agreeable modes of travelling for both, than labour had found out. He procured that fleet and untiring horse Exchange; and when he proposed to Labour that they should mount together, he claimed the right, and kept it, for their mutual benefit, of taking the direction of the horse. For this reason, as it appears to us, we are called upon to assign to one of the companions, according to the practice of the old Knights Templars, the privilege of sitting before the other—holding the reins, indeed, but in all respects having a community of interests, and an equality of duties, as well as rights, with his fellow-traveller.

Let us endeavour to advance another step in the illustration of these positions, by going back to the prodigal who had spent all his substance. Let us survey him at the moment when he had made the wise, and in many respects heroic, resolution to pass from the condition of a consumer into that of a producer. The story says, "The first thing that drew his attention was a heap of coals shot out of a cart on the pavement before a house. He offered himself to shovel or wheel them into the place where they were to be laid, and was employed." Here, then, we see that the labour of this man was wholly and imperatively directed by accumulation. It was directed as absolutely by the accumulation of others as the labour of Dampier's Moskito Indian was directed by his own accumulation. The Indian could not labour profitably—he could not obtain fish and goats for his food, instead of seal's flesh—till he had called into action the power which he possessed in his knife and his gun-barrel. The prodigal had no accumulation whatever of his own. He had not even the accumulation of peculiar skill in any mode of labour;—for a continual process of waste enlarges neither the mental nor physical faculties, and generally leaves the wretched being who has to pass into the new condition of a producer as helpless as the weakest child. He had nothing but the lowest power, of labouring without peculiar knowledge or skill. He had, however, an intensity and consistency of purpose which raised this humble power into real strength. He was determined never to go backward—always to go on. He knew, too, his duties as well as his rights; and he saw that he must wholly accommodate his power to the greater power which was in action around him. When he passed into the condition of a producer, he saw that his powers and rights were wholly limited and directed by the principles necessary to advance production; and that his own share of what he assisted in producing must be measured by the laws which enabled him to produce at all. He found himself in a position where his labour was absolutely governed by the system of exchanges. No other system could operate around him, because he was in a civilized country. Had he been thrown upon a desert land without food and shelter, his labour must have been instantly and directly applied to procuring food and shelter. He was equally without food and shelter in a civilized country. But the system of exchanges being in action, he did not apply his labour directly to the production of food and lodging for himself. He added by his labour a new value to a heap of coals; he enabled another man more readily to acquire the means of warmth; and by this service, which he exchanged for "a few pence" and "a small gratuity of meat and drink," he indirectly obtained food and lodging. He conferred an additional value upon a heap of coals; and that additional value was represented by the "few pence" and "a small gratuity of meat and drink." Had the system of exchange been less advanced, that is, had society been less civilized, he would probably have exchanged his labour for some object of utility, by another and a ruder mode. He would have received a portion of the coals as the price of the labour by which he gave an additional value to the whole heap. But mark the inconvenience of such a mode of exchange. His first want was food; his next, shelter: had he earned the coals, he must have carried them about till he had found some other person ready to exchange food and lodging for coals. Such an occurrence might have happened, but it would have been a lucky accident. He could find all persons ready to exchange food and clothes for money—because money was ready again to exchange for other articles of utility which they might require, and which they would more readily obtain by the money than by the food and clothes which our labourer had received for them. During the course of the unprofitable labour of waiting till he had found an exchanger who wanted coals, he might have perished. What then gave him the means of profitable labour, and furnished him with an article which every one was ready to receive in exchange for articles of immediate necessity? Capital in two forms. The heap of coals was capital. The coals represented a very great and various accumulation of former labour that had been employed in giving them value. The coals were altogether valueless till labour had been employed to raise them from the pit, and to convey them to the door of the man who was about to consume them. But with what various helps had this labour worked! Mere manual labour could have done little or nothing with the coals in the pit. Machines had raised them from the pit. Machines had transported them from the pit to the door of the consumer. They would have remained buried in the earth but for large accumulations of knowledge, and large accumulations of pecuniary wealth to set that knowledge in action by exchanging with it. The heap of coals represented all this accumulation; and it more immediately represented the Circulating Capital of consumable articles of utility, which had been paid in the shape of wages, at every stage of the labour exercised in raising the coals from the mine, and conveying them to the spot in which the prodigal found them laid. The coals had almost attained their highest value by a succession of labour; but one labour was still wanting to give them the highest value. They were at their lowest value when they remained unbroken in the coal-pit; they were at their highest value when they were deposited in the cellar of the consumer. For that last labour there was circulating capital ready to be exchanged. The man whose course of production we have been tracing imparted to them this last value, and for this labour he received a "few pence" and a "gratuity of meat and drink." These consumable commodities, and the money which might be exchanged for other consumable commodities, were circulating capital. They supplied his most pressing wants with incomparably more readiness and certainty than if he had been turned loose amongst the unappropriated productions of nature, with unlimited freedom and absolute rights. In the state in which he was actually placed his rights were limited by his duties,—but this balance of rights and duties was the chief instrument in the satisfaction of his wants. Let us examine the principle a little more in detail.

An exchange was to be carried on between the owner of the coals and the man who was willing to shovel them into the owner's cellar. The labourer did not want any distinct portion of the coals, but he wanted some articles of more urgent necessity in exchange for the new value which he was ready to bestow upon the coals. The object of each exchanger was, that labour should be exchanged with capital. That object could not have been accomplished, or it would have been accomplished slowly, imperfectly, and therefore unprofitably, unless there had been interchangeable freedom and security for both exchangers,—for the exchanger of capital, and the exchanger of labour. The first right of the labourer was, that his labour should be free;—the first right of the capitalist was, that his capital should be free. The rights of each were built upon the security of property. Could this security have been violated, it might have happened, either that the labourer should have been compelled to shovel in the coals—or, that the capitalist should have been compelled to employ the labourer to shovel them in. Had the lot of the unfortunate prodigal been cast in such a state of society as would have allowed this violation of the natural rights of the labourer and the capitalist, he would have found little accumulation to give a profitable direction to his labour. He would have found production suspended, or languishing. There would probably have been no heap of coals wanting his labour to give them the last value;—for the engines would have been idle that raised them from the pit, and the men would have been idle that directed the engines. The circulating capital that found wages for the men, and fuel for the engines, would have been idle, because it could not have worked with security. Accumulation, therefore, would have been suspended;—and all profitable labour would, in consequence, have been suspended. It was the unquestionable right of the labourer that his labour should be free; but it was balanced by the right of the capitalist that his accumulation should be secure. Could the labour have seized upon the capital, or the capital upon the labour, production would have been stopped altogether, or in part. The mutual freedom and security of labour and capital compel production to go forward; and labour and capital take their respective stations, and perform their respective duties, altogether with reference to the laws which govern production. These laws are founded upon the natural action of the system of exchange, carrying forward all its operations by the natural action of the great principle of demand and supply. When capital and labour know how to accommodate themselves to the direction of these natural laws, they are in a healthy state with respect to their individual rights, and the rights of industry generally. They are in that state in which each is working to the greatest profit in carrying forward the business of production.

The story of the prodigal goes on to say, "He then looked out for the next thing that might chance to offer; and went with indefatigable industry through a succession of servile employments, in different places, of longer and shorter duration." Here we see the principle of Demand and Supply still in active operation. "He *looked out* for the next thing that might chance to offer." He was ready with his supply of labour immediately that he saw a demand for it. Doubtless, the "indefatigable industry" with which he was ready with his supply created a demand, and thus he had in some degree a control over the demand. But in most cases the demand went before the supply, and he had thus to watch and wait upon the demand. In many instances demand and supply exercise a joint influence and control, each with regard to the other. Pliny, the Roman naturalist, relates that in the year 454 of the building of Rome (300 years before Christ) a number of barbers came over from Sicily to shave the Romans, who till that time had worn long beards. But the barbers came in consequence of being sent for by a man in authority. The demand here distinctly went before the supply; but the supply, doubtless, acted greatly upon the demand. During a time of wild financial speculation in Paris, created by what is called the Mississippi bubble, a hump-backed man went daily into the street where the stock-jobbers were accustomed to assemble, and earned money by allowing them to sign their contracts upon the natural desk with which he was encumbered. The hump-back was doubtless a shrewd fellow, and saw the difficulty under which the stock-jobbers laboured. He supplied what they appeared to want; and a demand was instantly created for his hump. He was well paid, says the story. That was because the supply was smaller than the demand. If other men with humps had been attracted by the demand, or if persons had come to the street with portable desks more convenient than the hump, the reward of his service would naturally have become less. He must have yielded to the inevitable law by which the amount of circulating capital, as compared with the number of labourers, prescribes the terms upon which capital and labour are united.

By following the direction which capital gave to his industry, the prodigal, whose course we have traced up to the point when he went into the condition of a labourer, became at length a capitalist. "He had gained, after a considerable time, money enough to purchase, in order to sell again, a few cattle, of which he had taken pains to understand the value. He speedily, but cautiously, turned his first gains into second advantages; retained, without a single deviation, his extreme parsimony; and thus advanced by degrees into larger transactions and incipient wealth. The final result was that he more than recovered his lost possessions, and died an inveterate miser, worth 60,000*l.*"

He gained "money," and he "purchased" cattle. In the simple transaction which has been recorded of the first exchange of the prodigal's labour for capital, we find the circumstances which represent every exchange of labour for capital. The prodigal wanted meat and drink, and he gave labour in exchange for meat and drink; the capitalist wanted the produce of labour —he wanted a new value bestowed upon his coals by labour—and he gave meat and drink in exchange for the labour which the prodigal had to give. But the prodigal wanted something beyond the meat and drink which was necessary for the supply of the day. He had other immediate necessities beyond food; and he had determined to accumulate capital. He therefore required "a few pence" in addition to the "meat and drink." The capitalist held that the labour performed had conferred a value upon his property, which would be fairly exchanged for the pence in addition to the food, and he gave, therefore, in exchange, that portion of his capital which was represented by the money and by the food. This blending of one sort of consumable commodity, and of the money which represented any other consumable commodity which the money could be exchanged with, was an accident arising out of the peculiar circumstances in which the prodigal happened to be placed. In ordinary cases he would have received the money alone,—that is, he would have received a larger sum of money to enable him to exchange for meat and drink, instead of receiving them in direct payment. It is clear, therefore, that as the money represented one portion of the consumable commodities which were ready to pay for the labour employed in giving a new value to the coals, it might represent another portion—the meat, for instance, without the drink; or it might represent all the consumable commodities, meat, drink, lodging, clothes, fuel, which that particular labourer might want; and even represent the accumulation which he might extract out of his self-denial as to the amount of meat, drink, lodging, clothes, and fuel which he might require as a consumer; and the farthing saved out of his money-payment might be the nest-egg which was to produce the increase out of which he purchased cattle, and died a rich miser.

We may be excused for calling attention to the fact, which is a very obvious one, that if the labourer, whose story we have told, had received a portion of the coals upon which he had conferred a new value in exchange for the labour which produced that value, he would have been paid in a way very unfavourable for production. It would have required a new labour before the coals could have procured him the meat, and drink, and lodging of which he had an instant want; and he therefore must have received a larger portion of coals to compensate for his new labour, or otherwise his labour must have been worse paid. There would have been unprofitable labour, whose loss must have fallen somewhere,—either upon the capitalist or the labourer in the first instance, but upon both ultimately, because there would have been less production. All the unprofitable labour employed in bringing the exchange of the first labour for capital to maturity would have been so much power withdrawn from the efficiency of the next labour to be performed; and therefore production would have been impeded to the extent of that unprofitable labour. The same thing would have happened if, advancing a step forward in the science of exchange, the labourer had received an entire payment in meat and drink, instead of a portion of the coals, which he could have exchanged for meat and drink. Wanting lodging, he would have had to seek a person who wanted meat and drink in exchange for lodging, before he could have obtained lodging. But he had a few pence,—he had money. He had a commodity to exchange that he might divide and subdivide as long as he pleased, whilst he was carrying on an exchange,—that is, he might obtain as much lodging as he required for an equivalent portion of his money. If he kept his money, it would not injure by keeping as the food would. He might carry it from place to place more easily than he could carry the food. He would have a commodity to exchange, whose value could not be made matter of dispute, as the value of meat and drink would unquestionably have been. This commodity would represent the same value, with little variation, whether he kept it a day, or a week, or a month, or a year; and therefore would be the only commodity whose retention would advance his design of accumulating capital with certainty and steadiness. It is evident that a commodity possessing all these advantages must have some intrinsic qualities which all exchangers would recognise—that it must be a standard of value—at once a commodity possessing real value, and a measure of all other values. This commodity exists in all commercial or exchanging nations in the shape of coined metal. The metal itself possesses a real value, which represents the labour employed in producing it; and, in the shape of coin, represents also a measure of other value, because the value of the coin has been determined by the sanction of some authority which all admit. That authority is most conveniently expressed by a Government, as the representative of the aggregate power of society. The metal itself, unless in the shape of coined money, would not represent a definite value; because the metal might be depreciated in value by the admixture of baser or inferior metals, unless it bore the impress of authority to determine its value. The exchangers of the metal for other articles of utility could not, without great loss of labour, be constantly employed in reducing it to the test of value, even if they had the knowledge requisite for so ascertaining its value. It used to cost 1000*l*. a year to the Bank of England for the wages of those who weighed the gold coin brought to the Bank; and it has been estimated that 30,000 sovereigns pass over the Bank counter daily. A machine is now used at the Bank, which separates the full-weight sovereigns and the light ones, at the rate of 10,000 an hour. In Greece a piece of gold in the rude times was stamped with the figure of an ox, to indicate that it would exchange for an ox. In modern England, a piece of gold, called a sovereign, represents a certain weight in gold uncoined, and the Government stamp indicates its purity; whilst the perpetual separation of the light sovereigns from those of full-weight affords a security that very few light ones are in general circulation. A sovereign purchases so many pounds weight of an ox, and a whole ox purchases so many sovereigns. The great use of the coined metal is to save labour in exchanging the ox for other commodities. The money purchases the ox, and a portion of the ox again purchases some other commodity, such as a loaf of bread from the baker, who obtains a portion of the ox through the medium of the money, which is a standard of value between the bread and the beef. Our great poetical satirist, Pope, in conducting his invective against the private avarice and political corruption of his day, imagines a state of things in which, money and credit being abolished, ministers would bribe and be bribed in kind. It is a true picture of what would be universal, if the exchanges of men resolved themselves into barter:

> "A statesman's slumbers how this speech would spoil!
> 'Sir, Spain has sent a thousand jars of oil;
> Huge bales of British cloth blockade the door;
> A hundred oxen at your levee roar.'"

CHAPTER VI.

Importance of capital to the profitable employment of labour—Contrast between the prodigal and the prudent man: the Dukes of Buckingham and Bridgewater—Making good for trade—Unprofitable consumption—War against capital in the middle ages—Evils of corporate privileges—Condition of the people under Henry VIII.

If we have succeeded in making our meaning clear, by stating a general truth, not in an abstract form, but as brought out by various instances of the modes in which it is exhibited, we shall have led the reader to the conclusion that accumulation, or capital, is absolutely essential to the profitable employment of labour; and that the greater the accumulation the greater the extent of that profitable employment. This truth, however, has been denied altogether by some speculative writers;—and, what is more important, has been practically denied by the conduct of nations and individuals in the earlier state of society,—and is still denied by existing prejudices, derived from the current maxims of former days of ignorance and half-knowledge. With the speculative writers we have little to do. When Rousseau, for instance, advises governments not to secure property to its possessors, but to deprive them of all means of accumulating, it is sufficient to know that the same writer advocated the savage state, in which there should be no property, in preference to the social, which is founded on appropriation. Knowing this, and being convinced that the savage state, even with imperfect appropriation, is one of extreme wretchedness, we may safely leave such opinions to work their own cure. For it is not likely that any individual, however disposed to think that accumulation is an evil, would desire, by destroying accumulation, to pass into the condition, described by John Tanner, of a constant encounter with hunger in its most terrific forms: and seeing, therefore, the fallacy of such an opinion, he will also see that, if he partially destroys accumulation, he equally impedes production, and equally destroys his share in the productive power of capital and labour working together for a common good in the social state.

But, without going the length of wishing to destroy capital, there are many who think that accumulation is a positive evil, and that consumption is a positive benefit; and, therefore, that economy is an evil, and waste a benefit.

The course of a prodigal man is by many still viewed with considerable admiration. He sits up all night in frantic riot—he consumes whatever can stimulate his satiated appetite—he is waited upon by a crowd of unproductive and equally riotous retainers—he breaks and destroys everything around him with an unsparing hand—he rides his horses to death in the most extravagant attempts to wrestle with time and space; and when he has spent all his substance in these excesses, and dies an outcast and a beggar, he is said to have been a hearty fellow, and to have "made good for trade." When, on the contrary, a man of fortune economizes his revenue—lives like a virtuous and reasonable being, whose first duty is the cultivation of his understanding—eats and drinks with regard to his health—keeps no more retainers than are sufficient for his proper comfort and decency—breaks and destroys nothing—has respect to the inferior animals, as well from motives of prudence as of mercy—and dies without a mortgage on his lands; *he* is said to have been a stingy fellow who did not know how to "circulate his money." To "circulate money," to "make good for trade," in the once common meaning of the terms, is for *one* to consume unprofitably what, if economized, would have stimulated production in a way that would have enabled *hundreds*, instead of one, to consume profitably. Let us offer two historical examples of these two opposite modes of making good for trade, and circulating money. The Duke of Buckingham, "having been possessed of about 50,000*l.* a year, died in 1687, in a remote inn in Yorkshire, reduced to the utmost misery."[12] After a life of the most wanton riot, which exhausted all his princely resources, he was left at the last hour, under circumstances which are well described in the following lines by Pope:—

"In the worst inn's worst room, with mat half hung,
The floors of plaster, and the walls of dung,
On once a flock bed, but repair'd with straw,
With tape-tied curtains never meant to draw,
The George and Garter dangling from that bed
Where tawdry yellow strove with dirty red;
Great Villiers lies....
No wit to flatter left of all his store,
No fool to laugh at, which he valued more,
There, victor of his health, of fortune, friends,
And fame, this lord of useless thousands ends."

Contrast the course of this unhappy man with that of the Duke of Bridgewater, who devoted his property to really "making good for trade," by constructing the great canals which connect Manchester with the coal countries and with Liverpool. The Duke of Buckingham lived in a round of sensual folly: the Duke of Bridgewater limited his personal expenditure to 400*l.* a-year, and devoted all the remaining portion of his revenues to the construction of a magnificent work of the highest public utility. The one supported a train of cooks and valets and horse-jockeys: the other called into action the labour of thousands, and employed in the direction of that labour the skill of Brindley, one of the greatest engineers that any country has produced. The one died without a penny, loaded with debt, leaving no trace behind him but the ruin which his waste had produced: the other bequeathed almost the largest property in Europe to his descendants, and opened a channel for industry which afforded, and still affords, employment to thousands.

Brindley.

When a mob amused themselves by breaking windows, as was once a common recreation on an illumination night, by way of showing the amount of popular intelligence, some were apt to say they have "made good for trade." Is it not evident that the capital which was represented by the unbroken windows was really so much destroyed of the national riches when the windows were broken?—for if the windows had remained unbroken, the capital would have remained to stimulate the production of some new object of utility. The glaziers, indeed, replaced the windows; but there having been a destruction of windows, there must have been a necessary retrenchment in some other outlay, that would have afforded benefit to the consumer. Doubtless, when the glazier is called into activity by a mob breaking windows, some other trade suffers; for the man who has to pay for the broken windows must retrench somewhere, and, if he has less to lay out, some other person has less to lay out. The glass-maker, probably, makes more glass at the moment; but he does so to exchange with the capital that would otherwise have gone to the maker of clothes or of furniture: and, there being an absolute destruction of the funds for the maintenance of labour, by an unnecessary destruction of what former labour has produced, trade generally is injured to the extent of the destruction. Some now say that a fire makes good for trade. The only difference of evil between the fire which destroys a house, and the mob which breaks the windows, is, that the fire absorbs capital for the maintenance of trade, or labour, in the proportion of a hundred to one when compared with the mob. Some say that war makes good for trade. The only difference of pecuniary evil (the moral evils admit of no comparison) between the fire and the war is, that the war absorbs capital for the maintenance of trade, or labour, in the proportion of a million to a hundred when compared with the fire. If the incessant energy of production were constantly repressed by mobs, and fires, and wars, the end would be that consumption would altogether exceed production; and that then the producers and the consumers would both be starved into wiser courses, and perceive that nothing makes good for trade but profitable industry and judicious expenditure. Prodigality devotes itself too much to the satisfaction of present wants: avarice postpones too much the present wants to the possible wants of the future. Real economy is the happy measure between the two extremes; and that only "makes good for trade," because, while it carries on a steady demand for industry, it accumulates a portion of the production of a country to stimulate new production. That judicious expenditure consists in

"The sense to value riches, with the art
T' enjoy them."

The fashion of "making good for trade" by unprofitable consumption is a relic of the barbarous ages. In the twelfth century a count of France commanded his vassals to plough up the soil round his castle, and he sowed the ground with coins of gold, to the amount of fifteen hundred guineas, that he might have all men talk of his magnificence. Piqued at the lordly prodigality of his neighbour, another noble ordered thirty of his most valuable horses to be tied to a stake and burnt alive, that he might exhibit a more striking instance of contempt for accumulation. In the latter part of the fourteenth century, a Scotch noble, Colin Campbell, on receiving a visit from the O'Neiles of Ireland, ostentatiously burnt down his house at Inverary upon their departure; and an Earl of Athol pursued the same course in 1528, after having entertained the papal legate, upon the pretence that it was "the constant habitude of the high-landers to set on fire in the morning the place which had lodged them the night before." When the feudal lords had so little respect for their own property, it was not likely that they would have much regard for the accumulation of others. The Jews, who were the great capitalists of the middle ages, and who really merit the gratitude of Europeans for their avarice, as that almost alone enabled any accumulation to go forward, and any production to increase, were, as it is well known, persecuted in every direction by the crown, by the nobles, by the people. When a solitary farmer or abbot attempted to accumulate corn, which accumulation could alone prevent the dreadful famines invariably resulting from having no stock that might be available upon a bad harvest, the people burnt the ricks of the provident men, by way of lessening the evils of scarcity. The consequence was, that no person thought of accumulating at all, and that the price of wheat often rose, just before the harvest, from five shillings a quarter to five pounds.

We are accustomed to read and talk of "merry England," but we sometimes fail to think how much real suffering lay beneath the surface of the merriment. Herrick, one of our charming old lyric poets, has sung the glories of the hock-cart—the cart that bore the full sheaves to the empty barn:—

>"The harvest swains and wenches bound
>For joy, to see the hock-cart crown'd;
>About the cart hear how the rout
>Of rural younglings raise the shout,
>Pressing before, some coming after,
>Those with a shout, and these with laughter.
>Some bless the cart, some kiss the sheaves,
>Some prank them up with oaken leaves;

> Some cross the fill-horse, some with great
> Devotion stroke the home-borne wheat."

Assuredly there was joy and there was devotion; for the laden cart made the difference between plenty and starvation. Before that harvest-home came there had been many an aching heart in the village hovels, for there was no store to equalize prices, and no communication to make the abundance of one district—much less of one country—mitigate the scarcity of another. It was not a question of the rise or the fall of a penny or two in the price of a loaf of bread; it was a question of bread or no bread.

The Hock-Cart.

During those dark periods the crown carried on the war against capital with an industry that could not be exceeded by that of the nobles or the people. Before the great charter the Commons complained that King Henry seized upon whatever was suited to his royal pleasure—horses, implements, food, anything that presented itself in the shape of accumulated labour. In the reign of Henry III. a statute was passed to remedy excessive distresses; from which it appeared that it was no unfrequent practice for the king's officers to take the opportunity of seizing the farmer's oxen at the moment when they were employed in ploughing, or, as the statute says, "winning the earth,"—taking them off, and starving them to death, or only restoring them with new and enormous exactions for their keep. Previous to the Charter of the Forest no man could dig a marl-pit on his own ground, lest the king's horses should fall into it when he was hunting. As late as the time of James I. we find, from a speech of the great Lord Bacon, that it was a pretty constant practice of the king's purveyors to extort large sums of money by threatening to cut down favourite trees which grew near a mansion-house or in avenues. Despotism, in all ages, has depopulated the finest countries, by rendering capital insecure, and therefore unproductive; insomuch that Montesquieu lays it down as a maxim, that lands are not cultivated in proportion to their fertility, but in proportion to their freedom. In the fifteenth century, in England, we find sums of money voted for the restoration of decayed towns and villages. Just laws would have restored them much more quickly and effectually. The state of agriculture was so low that the most absurd enactments were made to compel farmers to till and sow their own lands, and calling upon every man to plant at least forty beans. All the laws for the regulation of labourers, at the same period, assumed that they should be *compelled* to work, and not wander about the country,—just in the same way that farmers should be compelled to sow and till. It is perfectly clear that the towns would not have been depopulated, and gone to decay, if the accumulation of capital had not been obstructed by insecurity and wasted by ignorance, and that the same insecurity and the same waste rendered it necessary to assume that the farmer would not till and sow, and the labourer would not labour, without compulsion. The natural stimulus to industry was wanting, and therefore there was no industry, or only unprofitable industry. The towns decayed, the country was uncultivated—production languished—the people were all poor and wretched; and the government dreamt that acts of parliament and royal ordinances could rebuild the houses and cultivate the land, when the means of building and cultivation, namely, the capital of the country, was exhausted by injustice producing insecurity.

But if the king, the nobles, and the people of the middle ages conspired together, or acted at least as if they conspired, to prevent the accumulation of capital, the few capitalists themselves, by their monstrous regulations, which still throw some dark shadows over our own days, prevented that freedom of industry without which capital could not accumulate. Undoubtedly the commercial privileges of corporations originally offered some barriers against the injustice of the crown and of the nobility; but the good was always accompanied with an evil, which rendered it to a certain extent valueless. Where these privileges gave security, they were a good; where they destroyed freedom, they were an injury. Instead of encouraging the intercourse between one trade and another, they encircled every trade with the most absurd monopolies and exclusive privileges. Instead of rendering commerce free between one district and another, they, even in the same country, encompassed commerce with the most harassing restrictions, which separated county from county, and town from town, as if seas ran between them. If a man of Coventry came to London with his wares, he was encountered at every step with the privileges of companies; if the man of London sought to trade at Coventry, he was obstructed by the same corporate rights, preventing either the Londoner or the Coventry man trading with advantage. The revenues of every city were derived from forfeitures upon trades, almost all established upon the principle that, if one trade became too industrious or too clever, it would be the ruin of another trade. Every trade was fenced round with secrets; and the commonest trade, as we know from the language of an apprentice's indenture, was called an "art and mystery." All these follies went upon the presumption that "one man's gain is another man's loss," instead of vanishing before the truth, that, in proportion as the industry of all men is free, so will it be productive; and that production on all sides ensures a state of things in which every exchanger is a gainer, and no one a loser.

It is not to be wondered at that, while such opinions existed, the union of capital and labour should have been very imperfect; and that, while the capitalists oppressed the labourers, in the same way that they oppressed each other, the labourers should have thought it not unreasonable to plunder the capitalists. It is stated by Harrison, an old writer of credit,[13] that during the single reign of Henry VIII. seventy-two thousand thieves were hanged in England. No fact can exhibit in a stronger light the universal misery that must have existed in those days. The whole kingdom did not contain half a million grown-up males, so that, considering that the reign of Henry VIII. extended over two generations, about one man in ten must have been, to use the words of the same historian, "devoured and eaten up by the gallows." In the same reign the first statute against Egyptians (gipsies) was passed. These people went from place to place in great companies—spoke a cant language, which Harrison calls Pedler's French—and were subdivided into fifty-two different classes of thieves. The same race of people prevailed throughout Europe. Cervantes, the author of 'Don Quixote,' says of the Egyptians or Bohemians, that they seem to have been born for no other purpose than that of pillaging. While this universal plunder went forward, it is evident that the insecurity of property must have been so great that there could have been little accumulation, and therefore little production. Capital was destroyed on every side; and because profitable labour had become so scarce by the destruction of capital, one-half of the community sought to possess themselves of the few goods of the other half, not as exchangers but as robbers. As the robbers diminished the capital, the diminution of capital increased the number of robbers; and if the unconquerable energy of human industry had not gone on producing, slowly and painfully indeed, but still producing, the country would have soon fallen back to the state in which it was a thousand years before, when wolves abounded more than men. One great cause of all this plunder and misery was the oppression of the labourers.

[12] Ruffhead's Pope.

[13] Preface to the Chronicles of Holinshed.

CHAPTER VII.

Rights of labour—Effects of slavery on production—Condition of the Anglo-Saxons—Progress of freedom in England—Laws regulating labour—Wages and prices—Poor-law—Law of settlement.

Adam Smith, in his great work, 'The Wealth of Nations,' says, "The property which every man has in his own labour, as it is the original foundation of all other property, so it is the most sacred and inviolable. The patrimony of a poor man lies in the strength and dexterity of his hands; and to hinder him from employing this strength and dexterity in what manner he thinks proper, without injury to his neighbour, is a plain violation of this most sacred property." The right of property, in general, has been defined by another writer, M. Say, to be "the exclusive faculty guaranteed to a man, or body of men, to dispose, at their own pleasure, of that which belongs to them." There can be no doubt that labour is entitled to the same protection as a property that capital is entitled to. There can be no doubt that the labourer has rights over his labour which no government and no individual should presume to interfere with. There can be no doubt that, as an exchanger of labour for capital, the labourer ought to be assured that the exchange shall in all respects be as free as the exchanges of any other description of property. His rights as an exchanger are, that he shall not be compelled to part with his property, by any arbitrary enactments, without having as ample an equivalent as the general laws of exchange will afford him; that he shall be free to use every just means, either by himself or by union with others, to obtain such an equivalent; that he shall be at full liberty to offer that property in the best market that he can find, without being limited to any particular market; that he may give to that property every modification which it is capable of receiving from his own natural or acquired skill, without being narrowed to any one form of producing it. In other words, natural justice demands that the working-man shall work when he please, and be idle when he please, always providing that, if he make a contract to work, he shall not violate that engagement by remaining idle; that no labour shall be forced from him, and no rate of payment for that labour prescribed by statutes or ordinances; that he shall be free to obtain as high wages as he can possibly get, and unite with others to obtain them, always providing that in his union he does not violate that freedom of industry in others which is the foundation of his own attempts to improve his condition; that he may go from place to place to exchange his labour without being interfered with by corporate rights or monopolies of any sort, whether of masters or workmen; and that he may turn from one employment to the other, if he so think fit, without being confined to the trade he originally learnt, or may strike into any line of employment without having regularly learnt it at all. When the working-man has these rights secured to him by the sanction of the laws, and the concurrence of the institutions and customs of the country in which he lives, he is in the condition of a free exchanger. He has the full, uninterrupted, absolute possession of his property. He is upon a perfect legal equality with the capitalist. He may labour cheerfully with the well-founded assurance that his labour will be profitably exchanged for the goods which he desires for the satisfaction of his wants, as far as laws and institutions can so provide. In a word, he may assure himself that, if he possesses anything valuable to offer in exchange for capital, the capital will not be fenced round with any artificial barriers, or invested with any unnatural preponderance, to prevent the exchange being one of perfect equality, and therefore a real benefit to both exchangers.

We are approaching this desirable state in England. Indeed, there is scarcely any legal restriction acted upon which prevents the exchange of labour with capital being completely unembarrassed. Yet it is only within a few centuries that the working-men of this country have emerged from the condition of actual slaves into that of free labourers; it is only a few hundred years ago since the cultivator of the ground, the domestic servant, and sometimes even the artisan, was the absolute property of another man —bought, sold, let, without any will of his own, like an ox or a horse— producing nothing for himself—and transmitting the miseries of his lot to his children. The progress of civilization destroyed this monstrous system, in the same way that at the present day it is destroying it in Russia and other countries where slavery still exists. But it was by a very slow process that the English slave went forward to the complete enjoyment of the legal rights of a free exchanger. The transition exhibits very many years of gross injustice, of bitter suffering, of absurd and ineffectual violations of the natural rights of man; and of struggles between the capitalist and the labourer, for exclusive advantages, perpetuated by ignorant lawgivers, who could not see that the interest of all classes of producers is one and the same. We may not improperly devote a little space to the description of this dark and evil period. We shall see that while such a struggle goes forward—that is, while security of property and freedom of industry are not held as the interchangeable rights of the capitalist and the labourer—there can be little production and less accumulation. Wherever positive slavery exists—wherever the labourers are utterly deprived of their property in their labour, and are compelled to dispose of it without retaining any part of the character of voluntary exchangers—there are found idleness, ignorance, and unskilfulness; industry is enfeebled—the oppressor and the oppressed are both poor—there is no national accumulation. The existence of slavery amongst the nations of antiquity was a great impediment to their progress in the arts of life. The community, in such nations, was divided into a caste of nobles called citizens, and a caste of labourers called slaves. The Romans were rich, in the common sense of the word, because they plundered other nations; but they could not produce largely when the individual spirit to industry was wanting. The industry of the freemen was rapine: the slaves were the producers. No man will work willingly when he is to be utterly deprived of the power of disposing at his own will of the fruits of his labour; no man will work skilfully when the same scanty pittance is doled out to each and all, whatever be the difference in their talents and knowledge. Wherever the freedom of industry is thus violated, property cannot be secure. If Rome had encouraged free labourers, instead of breeding menial slaves, it could not have happened that the thieves, who were constantly hovering round the suburbs of the city, like vultures looking out for carrion, should have been so numerous that, during the insurrection of Catiline, they formed a large accession to his army. But Rome had to encounter a worse evil than that of the swarms of highwaymen who were ready to plunder whatever had been produced. Production itself was so feeble when carried on by the labour of slaves, that Columella, a writer on rural affairs, says the crops continued so gradually to fall off that there was a general opinion that the earth was growing old and losing its power of productiveness. Wherever slavery exists at the present day, there we find feeble production and national weakness. Poland, the most prolific corn-country in Europe, is unquestionably the poorest country. Poland has been partitioned, over and over again, by governments that knew her weakness; and she has been said to have fallen "without a crime." That is not correct. Her "crime" was, and is, the slavery of her labourers. There is no powerful class between the noble and the serf or slave; and whilst this state of things endures, Poland can never be independent, because she can never be industrious, and therefore never wealthy.

England, as we have said, once groaned under the evils of positive slavery. The Anglo-Saxons had what they called "live money," such as sheep and slaves. To this cause may be doubtless attributed the easy conquest of the country by the Norman invaders, and the oppression that succeeded that conquest. If the people had been free, no king could have swept away the entire population of a hundred thousand souls that dwelt in the country between the Humber and the Tees, and converted a district of sixty miles in length into a dreary desert, which remained for years without houses and without inhabitants. This the Conqueror did. In the reign of Henry II. the slaves of England were exported in large numbers to Ireland. These slaves, or villeins, as is the case in Russia and Poland at the present day, differed in the degree of the oppression which was exercised towards them. Some, called "villeins in gross," were at the absolute disposal of the lord—transferable from one owner to another, like a horse or a cow. Others, called "villeins regardant," were annexed to particular estates, and were called upon to perform whatever agricultural offices the lord should demand from them, not having the power of acquiring any property, and their only privilege being that they were irremoveable except with their own consent. These distinctions are not of much consequence, for, by a happy combination of circumstances, the bondmen of every kind, in the course of a century or two after the Conquest, were rapidly passing into the condition of free labourers. But still capital was accumulated so slowly, and labour was so unproductive, that the land did not produce the tenth part of a modern crop; and the country was constantly exposed to the severest inflictions of famine, whenever a worse than usual harvest occurred.

In the reign of Edward III. the woollen manufacture was introduced into England. It was at first carried on exclusively by foreigners; but as the trade extended, new hands were wanting, and the bondmen of the villages began to run away from their masters, and take refuge in the towns. If the slave could conceal himself successfully from the pursuit of his lord for a year and a day, he was held free for ever. The constant attraction of the bondmen to the towns, where they could work for hire, gradually emboldened those who remained as cultivators to assert their own natural rights. The nobility complained that the villeins refused to perform their accustomed services; and that corn remained uncut upon the ground. At length, in 1351, the 25th year of Edward III., the class of free labourers was first recognised by the legislature; and a statute was passed, oppressive indeed, and impolitic, but distinctly acknowledging the right of the labourer to assume the character of a free exchanger. Slavery, in England, was not wholly abolished by statute till the time of Charles II.: it was attempted in vain to be abolished in 1526. As late as the year 1775, the colliers of Scotland were accounted *ascripti glebæ*—that is, as belonging to the estate or colliery where they were born and continued to work. It is not necessary for us further to notice the existence of villeinage or slavery in these kingdoms. Our business is with the slow progress of the establishment of the rights of free labourers—and this principally to show that, during the long period when a contest was going forward between the capitalists and the labourers, industry was comparatively unproductive. It was not so unproductive, indeed, as during the period of absolute slavery; but as long as any man was compelled to work, or to continue at work, or to receive a fixed price, or to remain in one place, or to follow one employment, labour could not be held to be free—property could not be held to be secure—capital and labour could not have cordially united for production—accumulation could not have been certain and rapid.

In the year 1349 there was a dreadful pestilence in England, which swept off large numbers of the people. Those of the labourers that remained, following the natural course of the great principle of demand and supply, refused to serve, unless they were paid double the wages which they had received five years before. Then came the 'Statute of Labourers,' of 1351, to regulate wages; and this statute enacted what should be paid to haymakers, and reapers, and thrashers; to carpenters, and masons, and tilers, and plasterers. No person was to quit his own village, if he could get work at these wages; and labourers and artificers flying from one district to another in consequence of these regulations, were to be imprisoned.

Good laws, it has been said, execute themselves. When legislators make bad laws, there requires a constant increase of vigilance and severity, and constant attempts at reconciling impossibilities, to allow such laws to work at all. In 1360 the Statute of Labourers was confirmed with new penalties, such as burning in the forehead with the letter F those workmen who left their usual abodes. Having controlled the wages of industry, the next step was for these blind lawgivers to determine how the workmen should spend their scanty pittance; and accordingly, in 1363, a statute was passed to compel workmen and all persons not worth forty shillings to wear the coarsest cloth called russet, and to be served once a day with meat, or fish, and the offal of other victuals. We were not without imitations of such absurdities in other nations. An ordinance of the King of France, in 1461, determined that good and fat meat should be sold to the rich, while the poor should be allowed only to buy the lean and stinking.

While the wages of labour were fixed by statute, the price of wheat was constantly undergoing the most extraordinary fluctuations, ranging from 2*s.* a quarter to 1*l.* 6*s.* 8*d.* It was perfectly impossible that any profitable industry could go forward in the face of such unjust and ridiculous laws. In 1376 the Commons complained that masters were *obliged* to give their servants higher wages to prevent their running away; and that the country was covered with *staf-strikers* and *sturdy rogues*, who robbed in every direction. The villages were deserted by the labourers resorting to the towns, where commerce knew how to evade the destroying regulations of the statutes; and to prevent the total decay of agriculture, labourers were not allowed to move from place to place without letters patent:—any labourer, not producing such a letter, was to be imprisoned and put in the stocks. If a lad had been brought up to the plough till he was twelve years of age, he was compelled to continue in husbandry all his life; and in 1406 it was enacted that all children of parents not possessed of land should be brought up in the occupation of their parents. While the legislature, however, was passing these abominable laws, it was most effectually preparing the means for their extermination. Children were allowed to be sent to school in any part of the kingdom. When the light of education dawned upon the people, they could not long remain in the "darkness visible" that succeeded the night of slavery.

When the industry of the country was nearly annihilated by the laws regulating wages, it was found out that something like a balance should be preserved between wages and prices; and the magistrates were therefore empowered twice a year to make proclamation, according to the price of provisions, how much every workman should receive. The system, however, would not work well. In 1496 a new statute of wages was passed, the preamble of which recited that the former statutes had not been executed, because "the remedy by the said statutes is not very perfect." Then came a new remedy: that is, a new scale of wages for all trades; regulations for the hours of work and of rest; and penalties to prevent labour being transported from one district to another. As a necessary consequence of a fixed scale for wages, came another fixed scale for regulating the prices of provisions; till at last, in the reign of Henry VIII., lawgivers began to open their eyes to the folly of their proceedings, and the preamble of a statute says "that dearth, scarcity, good cheap, and plenty of cheese, butter, capons, hens, chickens, and other victuals necessary for man's sustenance, happeneth, riseth, and chanceth, of so many and divers occasions, that it is very hard and difficile to put any certain prices to any such things." Yet they went on with new scales, in spite of the hardness of the task; till at last some of the worst of these absurd laws were swept from the statute-book. The justices, whose principal occupation was to balance the scale of wages and labour, complained incessantly of the difficulty of the attempt; and the statute of the 5th Elizabeth acknowledged that these old laws "could not be carried into execution without the great grief and burden of the poor labourer and hired man." Still new laws were enacted to prevent the freedom of industry working out plenty for capitalists as well as labourers; and at length, in 1601, a general assessment was directed for the support of the impotent poor, and for setting the unemployed poor to work. The capitalists at length paid a grievous penalty for their two centuries of oppression; and had to maintain a host of paupers, that had gradually filled the land during these unnatural contests. It would be perhaps incorrect to say, that these contests alone produced the paupers that required this legislative protection in the reign of Elizabeth; but certainly the number of those paupers would have been far less, if the laws of industry had taken their healthy and natural course,—if capital and labour had gone hand in hand to produce abundance for all, and fairly to distribute that abundance in the form of profits and wages, justly balanced by the steady operation of demand and supply in a free and extensive market.

The whole of these absurd and iniquitous laws, which had succeeded the more wicked laws of absolute slavery, proceeded from a struggle on the part of the capitalists in land against the growing power and energy of free labour. If the capitalists had rightly understood their interests, they would have seen that the increased production of a thriving and happy peasantry would have amply compensated them for all the increase of wages to which they were compelled to submit; and that at every step by which the condition of their labourers was improved their own condition was also improved. If then capital had worked naturally and honestly for the encouragement of labour, there would have been no lack of labourers; and it would not have been necessary to pass laws to compel artificers, under the penalty of the stocks, to assist in getting in the harvest (5 Eliz.). If the labourers in agriculture had been adequately paid, they would not have fled to the towns; and if they had not been liable to cruel punishments for the exercise of this their natural right, the country would not have been covered with "valiant rogues and sturdy beggars."

The Law of Settlement, which, however modified, yet remains upon our statute-book, has been the curse of industry for nearly two centuries. All the best men of past times have cried out against its oppression. Roger North, soon after its enactment, in the time of Charles II., clearly enough showed its general operation:—"Where most work is, there are fewest people, and *è contra*. In Norfolk, Suffolk, and Essex, a labourer hath 12*d.* a day; in Oxfordshire, 8*d.*; in the North, 6*d.*, or less; and I have been credibly informed that in Cornwall a poor man will be thankful for 2*d.* a day and poor diet: and the value of provisions in all these places is much the same. Whence should the difference proceed? Even from plenty and scarcity of work and men, which happens crossgrainedly, so that one cannot come to the other." When men honestly went from home to seek work, they were called vagrants, and were confounded with the common beggars, for whom every severity was provided by the law—the stocks, the whip, the pillory, the brand. It was all in vain. Happy would it have been for the land if the law had left honest industry free, and in the case of dishonesty had applied itself to more effectual work than punishments and terror. One of our great judges, Sir Matthew Hale, said, long ago, what we even now too often forget—"The prevention of poverty, idleness, and a loose and disorderly education, even of poor children, would do more good to this kingdom than all the gibbets, and cauterizations, and whipping-posts, and jails in this kingdom." The whole scheme of legislation for the poor was to set the poor to work by forced contributions from capital. If the energy of the people had not found out how to set themselves to work in spite of bad laws, we might have remained a nation of slaves and paupers.

CHAPTER VIII.

Possessions of the different classes in England—Condition of Colchester in 1301—Tools, stock-in-trade, furniture, &c.—Supply of food—Comparative duration of human life—Want of facilities for commerce—Plenty and civilization not productive of effeminacy—Colchester in the present day.

It will be desirable to exhibit something like an average view of the extent of the possessions of all classes of society, and especially of the middling and labouring classes, in this country, at a period when the mutual rights of capitalists and labourers were so little understood as in the fourteenth century. We have shown how, at that time, there was a general round of oppression, resulting from ignorance of the proper interests of the productive classes; and it would be well also to show that during this period of disunion and contest between capital and labour, each plundering the other, and both plundered by arbitrary power, whether of the nobles or the crown, production went on very slowly and imperfectly, and that there was little to plunder and less to exchange. It is difficult to find the materials for such an inquiry. There is no very accurate record of the condition of the various classes of society before the invention of printing; and even after that invention we must be content to form our conclusions from a few scattered facts not recorded for any such purpose as we have in view, but to be gathered incidentally from slight observations which have come down to us. Yet enough remains to enable us to form a picture of tolerable accuracy; and in some points to establish conclusions which cannot be disputed. It is in the same way that our knowledge of the former state of the physical world must be derived from relics of that former state, to which the inquiries and comparisons of the present times have given an historical value. We know, for instance, that the animals of the southern countries once abounded in these islands, because we occasionally find their bones in quantities which could not have been accumulated unless such animals had been once native to these parts; and the remains of sea-shells upon the tops of hills now under the plough show us that even these heights have been heaved up from the bosom of the ocean. In the same way, although we have no complete picture of the state of property at the period to which we allude, we have evidence enough to describe that state from records which may be applied to this end, although preserved for a very different object.

In the reign of Edward III., Colchester, in Essex, was considered the tenth city in England in point of population. It then paid a poll-tax for 2955 lay persons. In 1301, about half a century before, the number of inhabitant housekeepers was 390; and the whole household furniture, utensils, clothes, money, cattle, corn, and every other property found in the town, was valued at 518*l.* 16*s.* 0-3/4*d.* This valuation took place on occasion of a subsidy or tax to the crown, to carry on a war against France; and the particulars, which are preserved in the Rolls of Parliament, exhibit with great minuteness the classes of persons then inhabiting that town, and the sort of property which each respectively possessed. The trades exercised in Colchester were the following:—baker, barber, blacksmith, bowyer, brewer, butcher, carpenter, carter, cobbler, cook, dyer, fisherman, fuller, furrier, girdler, glass-seller, glover, linen-draper, mercer and spice-seller, miller, mustard and vinegar seller, old clothes seller, saddler, tailor, tanner, tiler, weaver, wood-cutter, and wool-comber. If we look at a small town of the present day, where such a variety of occupations are carried on, we shall find that each tradesman has a considerable stock of commodities, abundance of furniture and utensils, clothes in plenty, some plate, books, and many articles of convenience and luxury to which the most wealthy dealers and mechanics of Colchester of the fourteenth century were utter strangers. That many places at that time were much poorer than Colchester there can be no doubt: for here we see the division of labour was pretty extensive, and that is always a proof that production is going forward, however imperfectly. We see, too, that the tradesmen were connected with manufactures in the ordinary use of the term; or there would not have been the dyer, the glover, the linen-draper, the tanner, the weaver, and the wool-comber. There must have been a demand for articles of foreign commerce, too, in this town, or we should not have had the spice-seller. Yet, with all these various occupations, indicating considerable profitable industry when compared with earlier stages in the history of this country, the whole stock of the town was valued at little more than 500*l.* Nor let it be supposed that this smallness of capital can be accounted for by the difference in the standard of money; although that difference is considerable. We may indeed satisfy ourselves of the small extent of the capital of individuals at that day, by referring to the inventory of the articles upon which the tax we have mentioned was laid at Colchester.

The whole stock of a carpenter's tools was valued at one shilling. They altogether consisted of two broad axes, an adze, a square, and a navegor or spoke-shave. Rough work must the carpenter have been able to perform with these humble instruments; but then let it be remembered that there was little capital to pay him for finer work, and that very little fine work was consequently required. The three hundred and ninety housekeepers of Colchester then lived in mud huts, with a rough door and no chimney. Harrison, speaking of the manners of a century later than the period we are describing, says, "There were very few chimneys even in capital towns: the fire was laid to the wall, and the smoke issued out at the roof, or door, or window. The houses were wattled, and plastered over with clay; and all the furniture and utensils were of wood. The people slept on straw pallets, with a log of wood for a pillow." When this old historian wrote, he mentions the erection of chimneys as a modern luxury. We had improved little upon our Anglo-Saxon ancestors in the article of chimneys. In their time Alcuin, an abbot who had ten thousand vassals, writes to the emperor at Rome that he preferred living in his smoky house to visiting the palaces of Italy. This was in the ninth century. Five hundred years had made little difference in the chimneys of Colchester. The nobility had hangings against the walls to keep out the wind, which crept in through the crevices which the builder's bungling art had left: the middle orders had no hangings. Shakspere alludes to this rough building of houses even in his time:—

"Imperial Cæsar, dead and turn'd to clay,
Might stop a hole to keep the wind away."

Even the nobility went without glass to their windows in the fourteenth and fifteenth centuries. "Of old time," says Harrison, "our country houses, instead of glass, did use much lattice, and that made either of wicker or fine rifts of oak, in checkerwise." When glass was introduced, it was for a long time so scarce that at Alnwick Castle, in 1567, the glass was ordered to be taken out of the windows, and laid up in safety, when the lord was absent.

The mercer's stock-in-trade at Colchester was much upon a level with the carpenter's tools. It was somewhat various, but very limited in quantity. The whole comprised a piece of woollen cloth, some silk and fine linen, flannel, silk purses, gloves, girdles, leather purses, and needle-work; and it was altogether valued at 3*l.* There appears to have been another dealer in cloth and linen in the town, whose store was equally scanty. We were not much improved in the use of linen a century later. We learn from the Earl of Northumberland's household book, whose family was large enough to consume one hundred and sixty gallons of mustard during the winter with their salt meat, that only seventy ells of linen were allowed for a year's consumption. In the fourteenth century none but the clergy and nobility wore white linen. As industry increased, and the cleanliness of the middle classes increased with it, the use of white linen became more general; but even at the end of the next century, when printing was invented, the paper-makers had the greatest difficulty in procuring rags for their manufacture; and so careful were the people of every class to preserve their linen, that night-clothes were never worn. Linen was so dear that Shakspere makes Falstaff's shirts eight shillings an ell. The more sumptuous articles of a mercer's stock were treasured in rich families from generation to generation; and even the wives of the nobility did not disdain to mention in their wills a particular article of clothing, which they left to the use of a daughter or a friend. The solitary old coat of a baker came into the Colchester valuation; nor is this to be wondered at, when we find that even the soldiers at the battle of Bannockburn, about this time, were described by an old rhymer as "well near all naked."

The household furniture found in use amongst the families of Colchester consisted, in the more wealthy, of an occasional bed, a brass pot, a brass cup, a gridiron, and a rug or two, and perhaps a towel. Of chairs and tables we hear nothing. We learn from the Chronicles of Brantôme, a French historian of these days, that even the nobility sat upon chests in which they kept their clothes and linen. Harrison, whose testimony we have already given to the poverty of these times, affirms, that if a man in seven years after marriage could purchase a flock bed, and a sack of chaff to rest his head upon, he thought himself as well lodged as the lord of the town, "who peradventure lay seldom on a bed entirely of feathers." An old tenure in England, before these times, binds the vassal to find straw even for the king's bed. The beds of flock, the few articles of furniture, the absence of chairs and tables, would have been of less consequence to the comfort and health of the people, if they had been clean; but cleanliness never exists without a certain possession of domestic conveniences. The people of England, in the days of which we are speaking, were not famed for their attention to this particular. Thomas à Becket was reputed extravagantly nice, because he had his parlour strewed every day with clean straw. As late as the reign of Henry VIII., Erasmus, a celebrated scholar of Holland, who visited England, complains that the nastiness of the people was the cause of the frequent plagues that destroyed them; and he says, "their floors are commonly of clay, strewed with rushes, under which lie unmolested a collection of beer, grease, fragments, bones, spittle, excrements of dogs and cats, and of everything that is nauseous." The elder Scaliger, another scholar who came to England, abuses the people for giving him no convenience to wash his hands. Glass vessels were scarce, and pottery was almost wholly unknown. The Earl of Northumberland, whom we have mentioned, breakfasted on trenchers and dined on pewter. While such universal slovenliness prevailed as Erasmus has described, it is not likely that much attention was generally paid to the cultivation of the mind. Before the invention of printing, at the time of the valuation of Colchester, books in manuscript, from their extreme costliness, could be purchased only by princes. The royal library of Paris, in 1378, consisted of nine hundred and nine volumes,—an extraordinary number. The same library now comprises upwards of four hundred thousand volumes. But it may fairly be assumed that, where one book could be obtained in the fourteenth century by persons of the working classes, four hundred thousand may be as easily obtained now. Here then was a privation which existed five hundred years ago, which debarred our ancestors from more profit and pleasure than the want of beds, and chairs, and linen; and probably, if this privation had continued, and men therefore had not cultivated their understandings, they would not have learnt to give any really profitable direction to their labour, and we should still have been as scantily supplied with furniture and clothes as the good people of Colchester of whom we have been speaking.

Let us see what accumulated supply, or capital, of food the inhabitants of England had five centuries ago. Possessions in cattle are the earliest riches of most countries. We have seen that cattle was called "live money;" and it is supposed that the word capital, which means stock generally, was derived from the Latin word "capita," or heads of beasts. The law-term "chattels" is also supposed to come from cattle. These circumstances show that cattle were the chief property of our ancestors. Vast herds of swine constituted the great provision for the support of the people; and these were principally fed, as they are even now in the New Forest, upon acorns and beech-mast. In Domesday Book, a valuation of the time of William the Conqueror, it is always mentioned how many hogs each estate can maintain. Hume the historian, in his Essays, alluding to the great herds of swine described by Polybius as existing in Italy and Greece, concludes that the country was thinly peopled and badly cultivated; and there can be no doubt that the same argument may be applied to England in the fourteenth century, although many swine were maintained in forests preserved for fuel. The hogs wandered about the country in a half-wild state, destroying, probably, more than they profitably consumed; and they were badly fed, if we may judge from a statute of 1402, which alleges the great decrease of fish in the Thames and other rivers, by the practice of feeding hogs with the fry caught at the weirs. The hogs' flesh of England was constantly salted for the winter's food. The people had little fodder for cattle in the winter, and therefore they only tasted fresh meat in the summer season. The mustard and vinegar seller formed a business at Colchester, to furnish a relish for the pork. Stocks of salted meat are mentioned in the inventory of many houses there, and live hogs as commonly. But salted flesh is not food to be eaten constantly, and with little vegetable food, without severe injury to the health. In the early part of the reign of Henry VIII., not a cabbage, carrot, turnip, or other edible root, grew in England. Two or three centuries before, certainly, the monasteries had gardens with a variety of vegetables; but nearly all the gardens of the laity were destroyed in the wars between the houses of York and Lancaster. Harrison speaks of wheaten bread as being chiefly used by the gentry for their own tables; and adds that the artificer and labourer are "driven to content themselves with horse-corn, beans, peason, oats, tares, and lentils." There is no doubt that the average duration of human life was at that period not one-half as long as at the present day. The constant use of salted meat, with little or no vegetable addition, doubtless contributed to the shortening of life, to say nothing of the large numbers constantly swept away by pestilence and famine. Till lemon-juice was used as a remedy for scurvy amongst our seamen, who also are compelled to eat salted meat without green vegetables, the destruction of life in the navy was something incredible. Admiral Hosier buried his ships' companies twice during a West India voyage in 1726, partly from the unhealthiness of the Spanish coast, but chiefly from the ravages of scurvy. Bad food and want of cleanliness swept away the people of the middle ages, by ravages upon their health that the limited medical skill of those days could never resist. Matthew Paris, an historian of that period, states that there were in his time twenty thousand hospitals for lepers in Europe.

The slow accumulation of capital in the early stages of the civilization of a country is in a great measure caused by the indisposition of the people to unite for a common good in public works, and the inability of governments to carry on these works, when their principal concern is war, foreign or domestic. The foundations of the civilization of this country were probably laid by our Roman conquerors, who carried roads through the island, and taught us how to cultivate our soil. Yet improvement went on so slowly that, even a hundred years after the Romans were settled here, the whole country was described as marshy. For centuries after the Romans made the Watling-street and a few other roads, one district was separated from another by the general want of these great means of communication. Bracton, a law-writer of the period we have been so constantly mentioning, holds that, if a man being at Oxford engage to pay money the same day in London, he shall be discharged of his contract, as he undertakes a physical impossibility. We find, as late as the time of Elizabeth, that her Majesty would not stay to breakfast at Cambridge because she had to travel twelve miles before she could come to the place, Hinchinbrook, where she desired to sleep. Where there were no roads, there could be few or no markets. An act of parliament of 1272 says that the religious houses should not be compelled to *sell* their provisions—a proof that there were no considerable stores except in the religious houses. The difficulty of navigation was so great, that William Longsword, son of Henry II., returning from France, was during three months tossed upon the sea before he could make a port in Cornwall. Looking, therefore, to the want of commerce proceeding from the want of communication—looking to the small stock of property accumulated to support labour—and looking, as we have previously done, to the incessant contests between the small capital and the misdirected labour, both feeble, because they worked without skill—we cannot be surprised that the poverty of which we have exhibited a faint picture should have endured for several centuries, and that the industry of our forefathers must have had a long and painful struggle before it could have bequeathed to us such magnificent accumulations as we now enjoy.

The writers who lived at the periods when Europe was slowly emerging from ignorance and poverty, through the first slight union of capital and labour as voluntary exchangers, complain of the increase of comforts as indications of the growing luxury and effeminacy of the people. Harrison says, "In times past men were contented to dwell in houses builded of sallow, willow, plum-tree, or elm; so that the use of oak was dedicated to churches, religious houses, princes' palaces, noblemen's lodgings, and navigation. But now, these are rejected, and nothing but oak any whit regarded. And yet see the change; for when our houses were builded of willow, then had we oaken men; but now that our houses are made of oak, our men are not only become willow, but many, through Persian delicacy crept in among us, altogether of straw, which is a sore alteration. In those days, the courage of the owner was a sufficient defence to keep the house in safety; but now, the assurance of the timber, double doors, locks, and bolts, must defend the man from robbing. Now have we many chimneys, and our tenderlings complain of rheums, catarrhs, and poses. Then had we none but rere-dosses, and our heads did never ache." These complaints go upon the same principle that made it a merit in Epictetus, the Greek philosopher, to have had no door to his hovel. We think he would have been a wiser man if he had contrived to have had a door. A story is told of a Highland chief, Sir Evan Cameron, that himself and a party of his followers being benighted, and compelled to sleep in the open air, when his son rolled up a ball of snow and laid his head upon it for a pillow, the rough old man kicked it away, exclaiming, "What, sir! are you turning effeminate?" We doubt whether Sir Evan Cameron and his men were braver than the English officers who fought at Waterloo; and yet many of these marched from the ball-room at Brussels in their holiday attire, and won the battle in silk stockings. It is an old notion that plenty of the necessaries and conveniences of life renders a nation feeble. We are told that the Carthaginian soldiers whom Hannibal carried into Italy were suddenly rendered effeminate by the abundance which they found around them at Capua. The commissariat of modern nations goes upon another principle; and believes that unless the soldier has plenty of food and clothing he will not fight with alacrity and steadiness. The half-starved soldiers of Henry V. won the battle of

Agincourt; but it was not because they were half-starved, but because they roused their native courage to cut their way out of the peril by which they were surrounded. The Russians of our time had a notion that the English could not fight on land, because for forty years we had been a commercial instead of a military nation. The battle of the Alma corrected their mistake. When we hear of ancient nations being enervated by abundance, we may be sure that the abundance was almost entirely devoured by a few tyrants, and that the bulk of the people were rendered weak by the destitution which resulted from the unnatural distribution of riches. We read of the luxury of the court of Persia—the pomp of the seraglios, and of the palaces—the lights, the music, the dancing, the perfumes, the silks, the gold, and the diamonds. The people are held to be effeminate. The Russians, from the hardy north, can lay the Persian monarchy any day at their feet. Is this national weakness caused by the excess of production amongst the people, giving them so extravagant a command over the necessaries and luxuries of life that they have nothing to do but drink of the full cup of enjoyment? Mr. Fraser, an English traveller, thus describes the appearance of a part of the country which he visited in 1821:—"The plain of Yezid-Khaust presented a truly lamentable picture of the general decline of prosperity in Persia. Ruins of large villages thickly scattered about with the skeleton-like walls of caravansaries and gardens, all telling of better times, stood like *memento moris* (remembrances of death) to kingdoms and governments; and the whole plain was dotted over with small mounds, which indicate the course of cannauts (artificial streams for watering the soil), once the source of riches and fertility, now all choked up and dry; for there is neither man nor cultivation to require their aid." Was it the luxury of the people which produced this decay—the increase of their means of production—their advancement in skill and capital; or some external cause which repressed production, and destroyed accumulation both of outward wealth and knowledge? "Such is the character of their rulers," says Mr. Fraser, "that the only measure of their demands is the power to extort on one hand, and the ability to give or retain on the other." Where such a system prevails, all accumulated labour is concealed, for it would otherwise be plundered. It does not freely and openly work to encourage new labour. Burckhardt, the traveller of Nubia, saw a farmer who had been plundered of everything by the pacha, because it came to the ears of the savage ruler that the unhappy man was in the habit of eating wheaten bread; and that, he thought, was too great a luxury for a subject. If such oppressions had not long ago been put down in England, we should still have been in the state of Colchester in the fourteenth century. When these iniquities prevailed, and there was neither freedom of industry nor security of property—when capital and labour were not united—when all men consequently worked unprofitably, because they worked without division of labour, accumulation of knowledge, and union of forces—there was universal poverty, because there was feeble production. Slow and painful were the steps which capital and labour had to make before they could emerge, even in part, from this feeble and degraded state. But that they have made a wonderful advance in five hundred years will not be difficult to show. It may assist us in this view if we compare the Colchester of the nineteenth century with the Colchester of the fourteenth, in a few particulars.

In the reign of Edward III. Colchester numbered 359 houses of mud, without chimneys, and with latticed windows. In the reign of Queen Victoria, according to the census of 1851, it has 4145 inhabited houses, containing a population of 19,443 males and females. The houses of the better class, those rented at ten pounds a year and upwards, are commonly built of brick, and slated or tiled; secured against wind and weather; with glazed windows and with chimneys; and generally well ventilated. The worst of these houses are supplied, as fixtures, with a great number of conveniences, such as grates, and cupboards, and fastenings. To many of such houses gardens are attached, wherein are raised vegetables and fruits that kings could not command two centuries ago. Houses such as these are composed of several rooms—not of one room only, where the people are compelled to eat and sleep and perform every office, perhaps in company with pigs and cattle—but of a kitchen, and often a parlour, and several bedrooms. These rooms are furnished with tables, and chairs, and beds, and cooking-utensils. There is ordinarily, too, something for ornament and something for instruction;—a piece or two of china, silver spoons, books, and not unfrequently a watch or clock. The useful pottery is abundant and of really elegant forms and colours; drinking-vessels of glass are universal. The inhabitants are not scantily supplied with clothes. The females are decently dressed, having a constant change of linen, and gowns of various patterns and degrees of fineness. Some, even of the humbler classes, are not thought to exceed the proper appearance of their station if they wear silk.

The men have decent working habits, strong shoes and hats, and a respectable suit for Sundays, of cloth often as good as is worn by the highest in the land. Every one is clean; for no house above the few hovels which still deform the country is without soap and bowls for washing, and it is the business of the females to take care that the linen of the family is constantly washed. The children, very generally, receive instruction in some public establishment; and when the labour of the day is over, the father thinks the time unprofitably spent unless he burns a candle to enable him to read a book or the newspaper. The food which is ordinarily consumed is of the best quality. Wheaten bread is no longer confined to the rich; animal food is not necessarily salted, and salt meat is used principally as a variety; vegetables of many sorts are plenteous in every market, and these by a succession of care are brought to higher perfection than in the countries of more genial climate from which we have imported them; the productions too of distant regions, such as spices, and coffee, and tea, and sugar, are universally consumed almost by the humblest in the land. Fuel, also, of the best quality, is abundant and comparatively cheap.

If we look at the public conveniences of a modern English town, we shall find the same striking contrast. Water is brought not only into every street, but into every house; the dust and dirt of a family is regularly removed without bustle or unpleasantness; the streets are paved, and lighted at night; roads in the highest state of excellence connect the town with the whole kingdom, and by means of railroads a man can travel several hundred miles in a few hours, and more readily than he could ten miles in the old time; and canal and sea navigation transport the weightiest goods with the greatest facility from each district to the other, and from each town to the other, so that all are enabled to apply their industry to what is most profitable for each and all. Every man, therefore, may satisfy his wants, according to his means, at the least possible expense of the transport of commodities. Every tradesman has a stock ready to meet the demand; and thus the stock of a very moderately wealthy tradesman of the Colchester of the present day is worth more than all the stock of all the different trades that were carried on in the same place in the fourteenth century. The condition of a town like Colchester—a flourishing market-town in an agricultural district—offers a fair point of comparison with a town of the time of Edward III.

CHAPTER IX.

Certainty the stimulus to industry—Effects of insecurity—Instances of unprofitable labour—Former notions of commerce—National and class prejudices, and their remedy.

Two of the most terrific famines that are recorded in the history of the world occurred in Egypt—a country where there is greater production, with less labour, than is probably exhibited in any other region. The principal labourer in Egypt is the river Nile, whose periodical overflowings impart fertility to the thirsty soil, and produce in a few weeks that abundance which the labour of the husbandman might not hope to command if employed during the whole year. But the Nile is a workman that cannot be controlled and directed, even by capital, the great controller and director of all work. The influences of heat, and light, and air, are pretty equal in the same places. Where the climate is most genial, the cultivators have least labour to perform in winning the earth; where it is least genial, the cultivators have most labour. The increased labour balances the small natural productiveness. But the inundation of a great river cannot be depended upon like the light and heat of the sun. For two seasons the Nile refused to rise, and labour was not prepared to compensate for this refusal; the ground refused to produce; the people were starved.

We mention these famines of Egypt to show that *certainty* is the most encouraging stimulus to every operation of human industry. We know that production as invariably follows a right direction of labour, as day succeeds to night. We believe that it will be dark to-night and light again to-morrow, because we know the general laws which govern light and darkness, and because our experience shows us that those laws are constant and uniform. We know that if we plough, and manure, and sow the ground, a crop will come in due time, varying indeed in quantity according to the season, but still so constant upon an average of years, that we are justified in applying large accumulations and considerable labour to the production of this crop. It is this certainty that we have such a command of the productive powers of nature as will abundantly compensate us for the incessant labour of directing those forces, which has during a long course of industry heaped up our manifold accumulations, and which enables production annually to go forward to an extent which even half a century ago would have been thought impossible. The long succession of labour, which has covered this country with wealth, has been applied to the encouragement of the productive forces of nature, and the restraint of the destroying. No one can doubt that, the instant the labour of man ceases to direct those productive natural forces, the destroying forces immediately come into action. Take the most familiar instance—a cottage whose neat thatch was never broken, whose latticed windows were always entire, whose whitewashed walls were ever clean, round whose porch the honeysuckle was trained in regulated luxuriance, whose garden bore nothing but what the owner planted. Remove that owner. Shut up the cottage for a year, and leave the garden to itself. The thatched roof is torn off by the wind and devoured by mice, the windows are driven in by storms, the walls are soaked through with damp and are crumbling to ruin, the honeysuckle obstructs the entrance which it once adorned, the garden is covered with weeds which years of after-labour will have difficulty to destroy:—

> "It was a plot
> Of garden-ground run wild, its matted weeds
> Mark'd with the steps of those whom, as they pass'd,
> The gooseberry-trees that shot in long lank slips,
> Or currants, hanging from their leafless stems
> In scanty strings, had tempted to o'erleap
> The broken wall."

Apply this principle upon a large scale. Let the productive energy of a country be suspended through some great cause which prevents its labour continuing in a profitable direction. Let it be overrun by a conqueror, or plundered by domestic tyranny of any kind, so that capital ceases to work with security. The fields suddenly become infertile, the towns lose their inhabitants, the roads grow to be impassable, the canals are choked up, the rivers break down their banks, the sea itself swallows up the land. Shakspere, a great political reasoner as well as a great poet, has described such effects in that part of 'Henry V.' when the Duke of Burgundy exhorts the rival kings to peace:—

> "Let it not disgrace me,

If I demand, before this royal view,
What rub, or what impediment, there is,
Why that the naked, poor, and mangled peace,
Dear nurse of arts, plenties, and joyful births,
Should not, in this best garden of the world,
Our fertile France, put up her lovely visage?
Alas! she hath from France too long been chas'd;
And all her husbandry doth lie on heaps,
Corrupting in its own fertility.
Her vine, the merry cheerer of the heart,
Unpruned, dies; her hedges even-pleach'd,
Like prisoners wildly overgrown with hair
Put forth disorder'd twigs: her fallow leas,
The darnel, hemlock, and rank fumitory
Doth root upon; while that the coulter rusts,
That should deracinate such savagery:
The even mead, that erst brought sweetly forth
The freckled cowslip, burnet, and green clover,
Wanting the scythe, all uncorrected, rank,
Conceives by idleness; and nothing teems
But hateful docks, rough thistles, kecksies, burs,
Losing both beauty and utility:
And as our vineyards, fallows, meads, and hedges,
Defective in their natures, grow to wildness;
Even so our houses, and ourselves and children,
Have lost, or do not learn, for want of time,
The sciences that should become our country."

Dykes of Holland: destruction by bursting.

We have heard it said that Tenterden steeple was the cause of Goodwin Sands. The meaning of the saying is, that the capital which was appropriated to keep out the sea from that part of the Kentish coast was diverted to the building of Tenterden steeple; and there being no funds to keep out the sea, it washed over the land.[14] The Goodwin Sands remain to show that man must carry on a perpetual contest to keep in subjection the forces of nature, which, as is said of fire, one of the forces, are good servants but bad masters. But these examples show, also, that in the social state our control of the physical forces of nature depends upon the right control of our own moral forces. There was injustice, doubtless, in misappropriating the funds which restrained the sea from devouring the land. Till men know that they shall work with justice on every side, they work feebly and unprofitably. England did not begin to accumulate largely and rapidly till the rights both of the poor man and the rich were to a certain degree established—till industry was free and property secure. Our great dramatic poet has described this security as the best characteristic of the reign of Queen Elizabeth:—

"In her days every man shall eat in safety
Under his own vine what he plants."

"Under his own vine"

Shakspere derived his image from the Bible, where a state of security is frequently indicated by direct allusion to a man sitting under the shade of his own fig-tree or his own vine. In the days of Elizabeth, as compared with previous eras, there was safety, and a man might

"sing
The merry songs of peace to all his neighbours."

We have gone on constantly improving these blessings. But let any circumstances again arise which may be powerful enough to destroy, or even molest, the freedom of industry and the security of property, and we should work once more without certainty. The elements of prosperity would not be constant and uniform. We should work with the apprehension that some hurricane of tyranny, no matter from what power, would arise, which would sweep away accumulation. When that hurricane did not rise, we might have comparative abundance, like the people of Egypt during the inundation of the Nile. We then should have an inundation of tranquillity. But if the tranquillity were not present—if lawless violence stood in the place of justice and security—we should be like the people of Egypt when the Nile did not overflow. We should suffer the extremity of misery; and that possible extremity would produce an average misery, even if tranquillity did return, because security had not returned. We should, if this state of things long abided, by degrees go back to the condition of Colchester in the fourteenth century, and thence to the universal marsh of two thousand years ago. The place where London stands would be, as it once was, a wilderness for howling wolves. The few that produced would again produce laboriously and painfully, without skill and without division of labour, because without accumulation; and it would probably take another thousand years, if men again saw the absolute need of security, to re-create what security has accumulated for our present use.

From the moment that the industry of this country began to work with security, and capital and labour applied themselves in union—perhaps not a perfect union, but still in union—to the great business of production, they worked with less and less expenditure of unprofitable labour. They continued to labour more and more profitably, as they laboured with knowledge. The labour of all rude nations, and of all uncultivated individuals, is labour with ignorance. Peter the wild boy, whom we have already mentioned, could never be made to perceive the right direction of labour, because he could not trace it through its circuitous courses for the production of utility. He would work under control, but, if left to himself, he would not work profitably. Having been trusted to fill a cart with manure, he laboured with diligence till the work was accomplished; but no one being at hand to direct him, he set to work as diligently to unload the cart again. He thought, as too many think even now, that the good was in the labour, and not in the results of the labour. The same ignorance exhibits itself in the unprofitable labour and unprofitable application of capital, even of persons far removed beyond the half-idiocy of Peter the wild boy. In the thirteenth century many of the provinces of France were overrun with rats, and the people, instead of vigorously hunting the rats, were persuaded to carry on a process against them in the ecclesiastical courts; and there, after the cause of the injured people and the injuring rats was solemnly debated, the rats were declared cursed and excommunicated if they did not retire in six days. The historian does not add that the rats obeyed the injunction; and doubtless the farmers were less prepared to resort to the profitable labour of chasing them to death when they had paid the ecclesiastics for the unprofitable labour of their excommunication. There is a curious instance of unprofitable labour given in a book on the Coal Trade of Scotland, written as recently as 1812. The people of Edinburgh had a passion for buying their coals in immense lumps, and, to gratify this passion, the greatest care was taken not to break the coals in any of the operations of conveying them from the pit to the cellar of the consumer. A wall of coals was first built within the pit, another wall under the pit's mouth, another wall when they were raised from the pit, another wall in the waggon which conveyed them to the port where they were shipped, another wall in the hold of the ship, another wall in the cart which conveyed them to the consumer, and another wall in the consumer's cellar; and the result of these seven different buildings-up and takings-down was, that after the consumer had paid thirty per cent. more for these square masses of coal than for coal shovelled together in large and small pieces, his servant had daily to break the large coals to bits to enable him to make any use of them. It seems extraordinary that such waste of labour and capital should have existed amongst a highly acute and refined community within the last forty years. They, perhaps, thought they were making good for trade, and therefore submitted to the evil; while the Glasgow people, on the contrary, by saving thirty per cent. in their coals, had that thirty per cent. to bestow upon new enterprises of industry, and for new encouragements to labour.

The unprofitable applications of capital and labour which the early history of the civilization of every people has to record, and which, amongst many, have subsisted even whilst they held themselves at the height of refinement, have been fostered by the ignorance of the great, and even of the learned, as to the causes which, advancing production or retarding it, advanced or retarded their own interests, and the interests of all the community. Princes and statesmen, prelates and philosophers, were equally ignorant of

> "What makes a nation happy, and keeps it so;
> What ruins kingdoms, and lays cities flat."

It was enough for them to consume; they thought it beneath them to observe even, much less to assist in, the direction of production. This was ignorance as gross as that of Peter the wild boy, or the excommunication of rats. It has always been the fashion of ignorant greatness to despise the mechanical arts. The pride of the Chinese mandarins was to let their nails grow as long as their fingers, to show that they never worked. Even European nobles once sought the same absurd distinction. In France, under the old monarchy, no descendant of a nobleman could embark in trade without the highest disgrace; and the principle was so generally recognised as just, that a French writer, even as recently as 1758, reproaches the sons of the English nobility for the contrary practice, and asks, with an air of triumph, how can a man be fit to serve his country in Parliament after having meddled with such paltry concerns as those of commerce? Montesquieu, a writer in most respects of enlarged views, holds that it is beneath the dignity of governments to interfere with such trumpery things as the regulation of weights and measures. Society might have well spared the interference of governments with weights and measures if they had been content to leave all commerce equally free. But, in truth, the regulation of weights and measures is almost a solitary exception to the great principle which governments ought to practise, of not interfering, or interfering little, with commerce.

Louis XIV. did not waste more capital and labour by his ruinous wars, and by his covering France with fortifications and palaces, than by the perpetual interferences of himself and his predecessors with the freedom of trade, which compelled capital and labour to work unprofitably. The naturally slow progress of profitable industry is rendered more slow by the perpetual inclination of those in authority to divert industry from its natural and profitable channels. It was therefore wisely said by a committee of merchants to Colbert, the prime minister of France in the reign of Louis XIV., when he asked them what measures government could adopt to promote the interests of commerce,—"Let us alone, permit us quietly to manage our own business." It is undeniable that the interests of all are best promoted when each is left free to attend to his own interests, under the necessary social restraints which prevent him doing a positive injury to his neighbour. It is thus that agriculture and manufactures are essentially allied in their interests; that unrestrained commerce is equally essential to the real and permanent interests of agriculture and manufactures; that capital and labour are equally united in their interests, whether applied to agriculture, manufactures, or commerce; that the producer and the consumer are equally united in their most essential interest, which is, that there should be cheap production. While these principles are not understood at all, and while they are imperfectly understood, as they still are by many classes and individuals, there must be a vast deal of unprofitable expenditure of capital, a vast deal of unprofitable labour, a vast deal of bickering and heart-burning between individuals who ought to be united, and classes who ought to be united, and nations who ought to be united; and as long as it is not felt by all that their mutual rights are understood and will be respected, there is a feeling of insecurity which more or less affects the prosperity of all. The only remedy for these evils is the extension of knowledge. Louis XV. proclaimed to the French that the English were their "véritables ennemis," their true enemies. When knowledge is triumphant it will be found that there are no "véritables ennemis," either among nations, or classes, or individuals. The prejudices by which nations, classes, and individuals are led to believe that the interest of one is opposed to the interest of another, are, nine times out of ten, as utterly absurd as the reason which a Frenchman once gave for hating the English—which was, "that they poured melted butter on their roast veal;" and this was not more ridiculous than the old denunciation of the English against the French, that "they ate frogs, and wore wooden shoes." When the world is disabused of the belief that the wealth of one nation, class, or individual must be created by the loss of another's wealth, then, and then only, will all men steadily and harmoniously apply to produce and to enjoy—to acquire prosperity and happiness—lifting themselves to the possession of good

"By Reason's light, on Resolution's wings."

CHAPTER X.

Employment of machinery in manufactures and agriculture—Erroneous notions formerly prevalent on this subject—Its advantages to the labourer—Spade-husbandry—The *principle* of machinery—Machines and tools—Change in the condition of England consequent on the introduction of machinery—Modern New Zealanders and ancient Greeks—Hand-mills and water-mills.

One of the most striking effects of the want of knowledge producing disunions amongst mankind that are injurious to the interests of each and all, is the belief which still exists amongst many well-meaning but unreflecting persons, that the powers and arrangements which Capital has created and devised for the advancement of production are injurious to the great body of working-men in their character of producers. The great forces by which capital and labour now work,—forces which are gathering strength every day,—are accumulation of skill and division of employments. It will be for us to show that the applications of science to the manufacturing arts have the effect of ensuring cheap production and increased employment. These applications of science are principally displayed in the use of MACHINERY; and we shall endeavour to prove that, although individual labour may be partially displaced, or unsettled for a time, by the use of this cheaper and better power than unassisted manual labour, all are great gainers by the general use of that power. Through that power all principally possess, however poor they may be, many of the comforts which make the difference between man in a civilized and man in a savage state; and further, that, in consequence of machinery having rendered productions of all sorts cheaper, and therefore caused them to be more universally purchased, it has really increased the demand for that manual labour, which it appears to some, reasoning only from a few instances, it has a tendency to diminish.

In the year 1827 a Committee of the House of Commons was appointed to examine into the subject of Emigration. The first person examined before that Committee was Joseph Foster, a working weaver of Glasgow. He told the Committee that he and many others, who had formed themselves into a society, were in great distress; that numbers of them worked at the *hand-*

loom from eighteen to nineteen hours a day, and that their earnings, at the utmost, did not amount to more than seven shillings a-week, and that sometimes they were as low as four shillings. That twenty years before that time they could readily earn a pound a-week by the same industry; and that as *power-loom* weaving had increased, the distress of the hand-weavers also had increased in the same proportion. The Committee then put to Joseph Foster the following questions, and received the following answers:—

> "*Q.* Are the Committee to understand that you attribute the insufficiency of your remuneration for your labour to the introduction of machinery?
>
> "*A.* Yes.
>
> "*Q.* Do you consider, therefore, that the introduction of machinery is objectionable?
>
> "*A.* We do not. The weavers in general, of Glasgow and its vicinity, do not consider that machinery can or ought to be stopped, or put down. They know perfectly well that machinery must go on, that it will go on, and that it is impossible to stop it. They are aware that every implement of agriculture or manufacture is a portion of machinery, and, indeed, everything that goes beyond the teeth and nails (if I may use the expression) is a machine. I am authorized, by the majority of our society, to say that I speak their minds, as well as my own, in stating this."

It is worthy of note how the common sense of this working-man, a quarter of a century ago, saw clearly the great principle which overthrows, in the outset, all unreasoning hostility to machinery. Let us follow out his principle.

Amongst the many accounts which the newspapers in December, 1830, gave of the destruction of machinery by agricultural labourers, we observed that in the neighbourhood of Aylesbury a band of mistaken and unfortunate men destroyed all the machinery of many farms, *down even to the common drills*. The men conducted themselves, says the county newspaper, with civility; and such was their consideration, that they moved the machines out of the farm-yards, to prevent injury arising to the cattle from the nails and splinters that flew about while the machinery was being destroyed. They *could not make up their minds* as to the propriety of destroying a horse-churn, and therefore that machine was passed over.

A quarter of a century has made a remarkable difference in the feelings, even in the least informed, with regard to machinery. The majority of the people now know, as the weavers of Glasgow knew in 1827, that "machinery must go on, that it will go on, and that it is impossible to stop it." We therefore, adapting this volume to the improved times in which it is now published, think it needless to urge, as fully as we once did, any of the notions of the labourers of Aylesbury to their inevitable conclusions. It is sufficient briefly to show, that, if the labourers had been successful in their career, had broken all those ingenious implements which have aided in rendering British agriculture the most perfect in the world, they would not have advanced a single step in obtaining more employment, or being better paid.

We will suppose, then, that the farmer has yielded to this violence; that the violence has had the effect which it was meant to have upon him; and that he takes on all the hands which were out of employ, to thrash and winnow, to cut chaff, to plant with a dibber instead of with a drill, to do all the work, in fact, by the dearest mode instead of the cheapest. But he employs *just as many people as are absolutely necessary*, and no more, for getting his corn ready for market, and for preparing, in a slovenly way, for the seed-time. In a month or two the victorious destroyers find that not a single hand the more of them is really employed. And why not? There are no drainings going forward, the hedges and ditches are neglected, the dung-heap is not turned over, the chalk is not fetched from the pit; in fact, all those labours are neglected which belong to a state of agricultural industry which is brought to perfection. *The farmer has no funds to employ in such labours*; he is paying a great deal more than he paid before for the same, or a less, amount of work, because his labourers choose to do certain labours with rude tools instead of perfect ones.

We will imagine that this state of things continues till the next spring. All this while the price of grain has been rising. Many farmers have ceased to employ capital at all upon the land. The neat inventions, which enabled them to make a living out of their business, being destroyed, they have abandoned the business altogether. A day's work will now no longer purchase as much bread as before. The horse, it might be probably found out, was as great an enemy as the drill-plough; for, as a horse will do the field-work of six men, there must be six men employed, without doubt, instead of one horse. But how would the fact turn out? If the farmer still went on, in spite of all these losses and crosses, he might employ men in the place of horses, but not a single man more than the number that would work at the price of the keep of one horse. To do the work of each horse turned adrift, he would require six men; but he would only have about a shilling a-day to divide between these six—the amount which the horse consumed.

As the year advanced, and the harvest approached, it would be discovered that not one-tenth of the land was sown: for although the ploughs were gone, because the horses were turned off, and there was plenty of *labour* for those who chose to labour for its own sake, or at the price of horse-labour, this amazing employment for human hands, somehow or other, would not quite answer the purpose. It has been calculated that the power of horses, oxen, &c., employed in husbandry in Great Britain is ten times the amount of human power. If the human power insisted upon doing all the work with the worst tools, the certainty is that not even one-tenth of the land could be cultivated. Where, then, would all this madness end? In the starvation of the labourers themselves, even if they were allowed to eat up all they had produced by such imperfect means. They would be just in the condition of any other barbarous people, that were ignorant of the inventions that constitute the power of civilization. They would eat up the little corn which they raised themselves, and have nothing to give in exchange for clothes, and coals, and candles, and soap, and tea, and sugar, and all the many comforts which those who are even the worst off are not wholly deprived of.

All this may appear as an extreme statement; and certainly we believe that no such evils could have happened: for if the laws had been passive, the most ignorant of the labourers themselves would, if they had proceeded to carry their own principle much farther than they had done, see in their very excesses the real character of the folly and wickedness to which it had led, and would lead them. Why should the labourers of Aylesbury not have destroyed the harrows as well as the drills? Why leave a machine which separates the clods of the earth, and break one which puts seed into it? Why deliberate about a horse-churn, when they were resolved against a winnowing-machine? The truth is, these poor men perceived, even in the midst of their excesses, the gross deception of the reasons which induced them to commit them. Their motive was a natural, and, if lawfully expressed, a proper impatience, under a condition which had certainly many hardships, and those hardships in great part produced by the want of profitable labour. But in imputing those hardships to machinery, they were at once embarrassed when they came to draw distinctions between one sort of machine and another. This embarrassment decidedly shows that there were fearful mistakes at the bottom of their furious hostility to machinery.

It has been said, by persons whose opinions are worthy attention, that spade-husbandry is, in some cases, better than plough-husbandry;—that is, that the earth, under particular circumstances of soil and situation, may be more fitly prepared for the influences of the atmosphere by digging than by ploughing. It is not our business to enter into a consideration of this question. The growth of corn is a manufacture, in which man employs the chemical properties of the soil and of the air in conjunction with his own labour, aided by certain tools or machines, for the production of a crop; and that power, whether of chemistry or machinery,—whether of the salt, or the chalk, or the dung, or the guano, which he puts upon the earth, or the spade or the plough which he puts into it,—that power which does the work easiest is necessarily the best, *because it diminishes the cost of production*. If the plough does not do the work so well as the spade, it is a less perfect machine; but the less perfect machine may be preferred to the more perfect, because, taking other conditions into consideration, it is a cheaper machine. If the spade, applied in a peculiar manner by the strength and judgment of the man using it, more completely turns up the soil, breaks the clods, and removes the weeds than the plough, which receives one uniform direction from man with the assistance of other animal power, then the spade is a more perfect machine in its combination with human labour than the plough is, worked with a lesser degree of the same combination. But still it may be a machine which cannot be used with advantage to the producer, and is therefore not desirable for the consumer. All such questions must be determined by the cost of production; and that cost in agriculture is made up of the rent of land, the profit of capital, and the wages of labour—or the portions of the produce belonging to the landlord, the farmer, and the labourer. Where rent is high, as in the immediate neighbourhood of large towns, it is important to have the labour performed as carefully as possible, and the succession of crops stimulated to the utmost extent by manure and labour. It is economy to turn the soil to the greatest account, and the land is cultivated as a garden. Where rent is low, it is important to have the labour performed upon other principles, because one acre cultivated by hand may cost more than two cultivated by the plough; and though it may be the truest policy to carry the productiveness of the earth as far as possible, field cultivation and garden cultivation must have essential differences. In one case, the machine called a spade is used; in the other, the machine called a plough is employed. The use of the one or the other belongs to practical agriculture, and is a question only of relative cost.

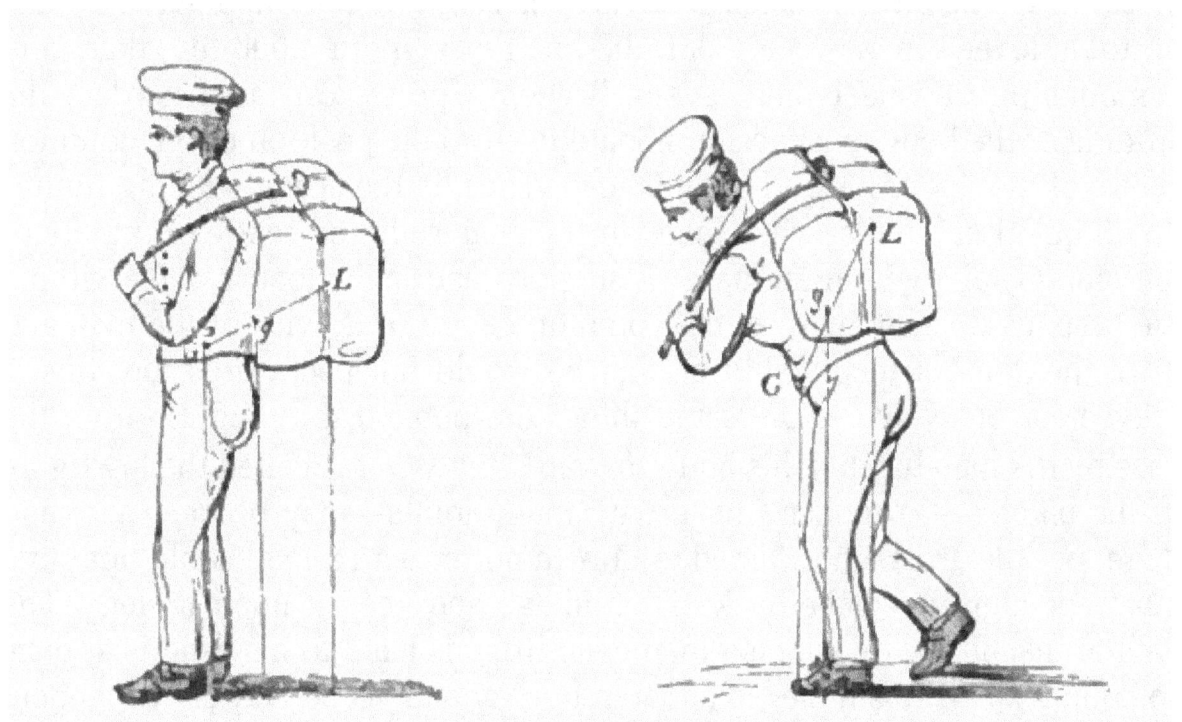

Centre of gravity

And this brings us to the great *principle* of all machinery. A tool of the simplest construction is a machine; a machine of the most curious construction is only a complicated tool. There are many cases in the arts, and there may be cases in agriculture, in which the human arm and hand, with or without a tool, may do work that no machine can so well perform. There are processes in polishing, and there is a process in copper-plate printing, in which no substance has been found to stand in the place of the human hand. And, if therefore the man with a spade alone does a certain agricultural work more completely than a man guiding a plough, and a team of horses dragging it (which we do not affirm or deny), the only reason for this is, that the man with the spade is a better machine than the man with the plough and the horses. The most stupid man that ever existed is, beyond all comparison, a machine more cunningly made by the hands of his Creator, more perfect in all his several parts, and with all his parts more exquisitely adapted to the regulated movement of the whole body, less liable to accidents, and less injured by wear and tear, than the most beautiful machine that ever was, or ever will be, invented. There is no possibility of supplying in many cases a substitute for the simplest movements of a man's body, by the most complicated movements of the most ingenious machinery. The laws of mechanism are the same whether applied to a man, or to a lever, or a wheel; but the man has more pliability than any combination of wheels and levers. "When a porter carries a burden, the attitude of the body must accommodate itself to the position of the common centre of gravity of himself and his load. Thus, in the accompanying figures it will be observed that when the man stands upright the centre of gravity of the man G falls within the base of support; and if his load L falls without the base, as does likewise g, the common centre of gravity of the man and load, the consequence would be that he would fall backwards; but this is prevented, or, which is the same thing, the point g is brought within the base by the man bending his body forward."[15] What is called the lay figure of the painter—a wooden image with many joints—may be bent here and there; but the artist who wanted to draw a porter with a load could not put a hundredweight upon the back of his image and keep it upon its legs. The natural machinery by which a man even lifts his hand to his head is at once so complex and so simple, so apparently easy and yet so entirely dependent upon the right adjustment of a great many contrary forces, that no automaton, or machine imitating the actions of man, could ever be made to effect this seemingly simple motion, without showing that the contrivance was imperfect,—that it was a mere imitation, and a very clumsy one. What an easy thing it appears to be for a farming man to thrash his corn with a flail; and yet what an expensive arrangement of wheels is necessary to produce the same effects with a thrashing-machine! The truth is, that the man's arm and the flail form a much more curious machine than the other machine of wheels, which does the same work; and the real question as regards the value of the two machines is, which machine in the greater degree lessens the cost of production?

We state this principle broadly, in our examination into the value of machinery in diminishing the cost of producing human food. A machine is not perfect because it is made of wheels or cylinders, employs the power of the screw or the lever, is driven by wind or water or steam, but because it best assists the labour of man, by calling into action some power which he does not possess in himself. If we could imagine a man entirely dispossessed of this power, we should see the feeblest of animal beings. He has no tools which are a part of himself, to build houses like the beaver, or cells like the bee. He has not even learnt from nature to build, instinctively, by certain and unchangeable rules. His power is in his mind; and that mind teaches him to subject all the physical world to his dominion, by availing himself of the forces which nature has spread around him. To act upon material objects he arms his weakness with tools and with machines. As we have before said, tools and machines are in principle the same. When we strike a nail upon the head with a hammer, we avail ourselves of a power which we find in nature—the effect produced by the concussion of two bodies; when we employ a water-wheel to beat out a lump of iron with a much larger hammer, we still avail ourselves of the same power. There is no difference in the nature of the instruments, although we call the one a tool, and the other a machine. Neither the tool nor the machine has any force of itself. In one case the force is in the arm, in the other in the weight of water which turns the wheel. The distinctions which have been taken between a tool and a machine are really so trivial, and the line of separation between the one and the other is so slight, that we can only speak of both as common instruments for adding to the efficiency of labour. The simplest application of a principle of mechanics to an every-day hand-tool may convert it into what is called a machine. Take a three-pronged fork—one of the universal tools;—fasten a rope to the end of the handle; put a log under the fork as a fulcrum; and we have a lever, when pulled down by the rope, which will grub up a strongly-rooted large shrub in a few minutes. The labourer has called in a powerful ally. The tool has become a machine.

A tool made a machine

The chief difference between man in a rude, and man in a civilized state of society is, that the one wastes his force, whether natural or acquired,—the other economizes, that is, saves it. The man in a rude state has very rude instruments; he therefore wastes his force: the man in a civilized state has very perfect ones; he therefore economizes it. Should we not laugh at the gardener who went to hoe his potatoes with a stick having a short crook at the end? It would be a tool, we should say, fit only for children to use. Yet such a tool was doubtless employed by some very ancient nations; for there is an old medal of Syracuse which represents this very tool. The common hoe of the English gardener is a much more perfect tool, because it saves labour. Could we have any doubt of the madness of the man who would propose that all iron hoes should be abolished, to furnish more extensive employ to labourers who should be provided only with a crooked stick cut out of a hedge? The truth is, if the working men of England had no better tools than crooked sticks, they would be in a state of actual starvation. One of the chiefs of New Zealand, before that country had been colonized by us, told Mr. Marsden, a missionary, that his wooden spades were all broken, and he had not an axe to make any more;—his canoes were all broken, and he had not a nail or a gimlet to mend them with;—his potato-grounds were uncultivated, and he had not a hoe to break them up with;—and that *for want of cultivation* he and his people would have nothing to eat. This shows the state of a people without tools. The man had seen English tools, and knew their value.

About three or four hundred years ago, from the times of king Henry IV. to those of king Henry VI., and, indeed, long before these reigns, there were often, as we have already mentioned, grievous famines in this country, because the land was very wretchedly cultivated. Men, women, and children perished of actual hunger by thousands; and those who survived kept themselves alive by eating the bark of trees, acorns, and pig-nuts. There were no machines then; but the condition of the labourers was so bad, that they could not be kept to work upon the land without those very severe and tyrannical laws noticed in Chap. VII., which absolutely forbade them to leave the station in which they were born as labourers, for any hope of bettering their condition in the towns. There were not labourers enough to till the ground, for they worked without any skill, with weak ploughs and awkward hoes. They were just as badly off as some of the people of Portugal and Spain, who are miserably poor, *because* they have bad machines; or as the Chinese labourers, who have scarcely any machines, and are the poorest in the world. There was plenty of labour to be performed, but the tools were so bad, and the want of agricultural knowledge so universal, that the land was never half cultivated, and therefore all classes were poorly off. They had little corn to exchange for manufactures, and in consequence the labourer was badly clothed, badly lodged, and had a very indifferent share of the scanty crop which he raised. The condition of the labourer would have proceeded from bad to worse, had agricultural improvement not gone forward to improve the general condition of all classes. Commons were enclosed; arable land was laid down to pasture; sheep were kept upon grass-land where wretched crops had before been growing. This was superseding labour to a great extent, and much clamour was raised about this plan, and probably a large amount of real distress was produced. But mark the consequence. Although the money wages of labour were lowered, because there were more labourers in the market, the real amount of wages was higher, for better food was created by pasturage at a cheaper rate. The labourer then got meat who had never tasted it before; and when the use of animal food became general, there were cattle and corn enough to be exchanged for manufactured goods, and the labourer got a coat and a pair of shoes, who had formerly gone half naked. Step by step have we been going in the same direction for two centuries; and the agricultural industry of Great Britain is now as much directed to the production of meat, milk, butter, cheese, as to the growth of corn and other cereals. The once simple husbandry of our forefathers has become a scientific manufacture.

Spinning.

There may be some persons still who object to machinery, because, having grown up surrounded with the benefits it has conferred upon them, without understanding the source of these benefits, they are something like the child who sees nothing but evil in a rainy day. The people of New Zealand very rarely came to us; but when they did come they were acute enough to perceive the advantages which machinery has conferred upon us, and the great distance in point of comfort between their state and ours, principally for the reason that they have no machinery, while we have a great deal. One of these men burst into tears when he saw a rope-walk; because he perceived the immense superiority which the process of spinning ropes gave us over his own countrymen. He was ingenious enough, and so were many of his fellow islanders, to have twisted threads together after a rude fashion; but he knew that he was a long way off from making a rope. The New Zealander saw the spinner in the rope-walk, moving away from a wheel, and gradually forming the hemp round his body into a strong cord. By the operation of the wheel he is enabled so to twine together a number of separate fibres, that no one fibre can be separated from the mass, but forms part of a hard and compact body. A series of these operations produces a cable, such as may hold a barge at anchor. The twisted fibres of hemp become yarn; many yarns become a strand; three strands make a rope; and three ropes make a cablet, or small cable. By carefully untwisting all its separate parts, the principle upon which it is constructed is evident. The operation is a complex one; and the rope-maker is a skilled workman. Rope-making machinery is now carried much farther. But the wheel that twisted the hemp into yarn was a prodigy to the inquiring savage.

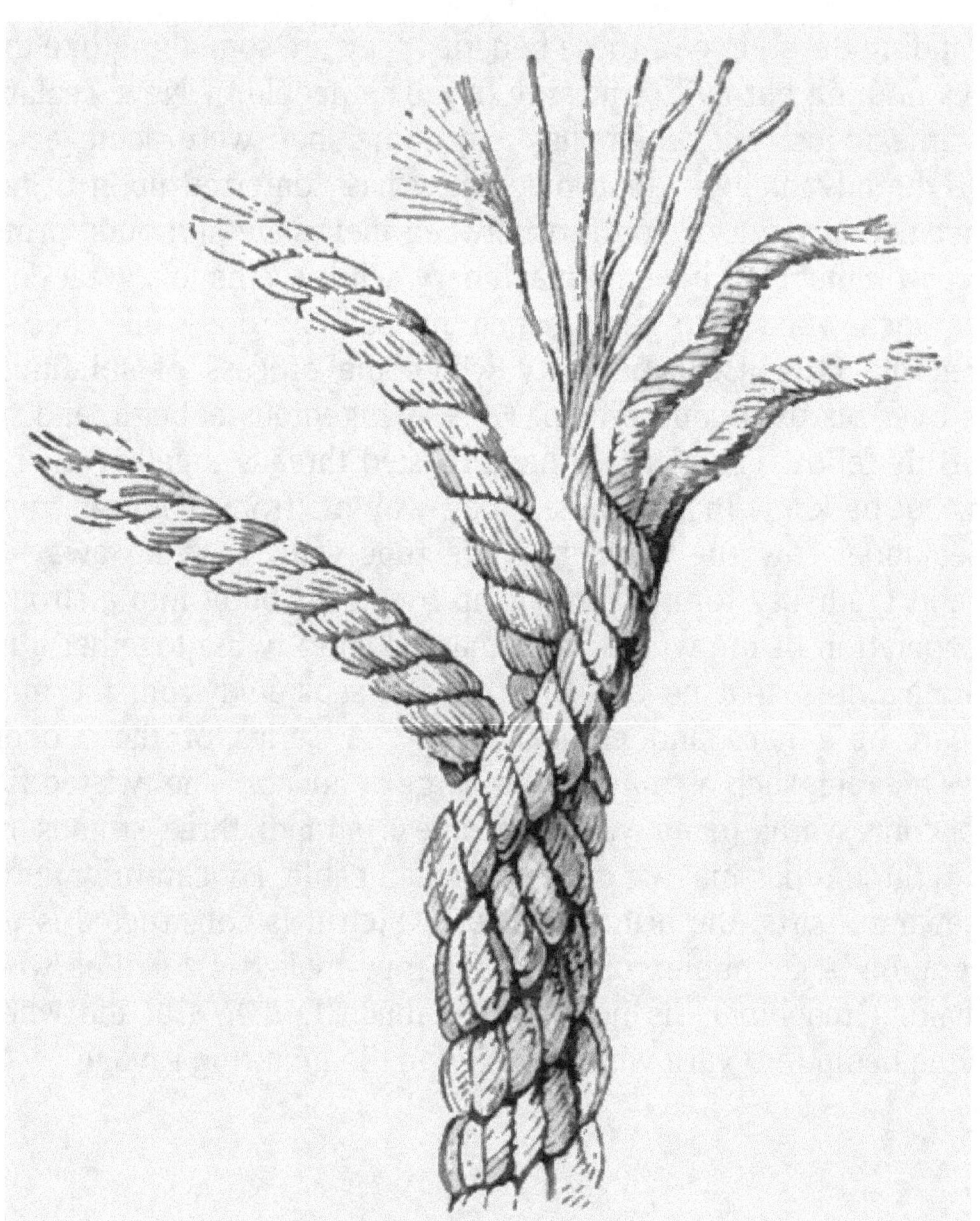

Analysis of a Cablet.

Another of these New Zealanders, and he was a very shrewd and intelligent person, carried back to his country a small hand-mill for grinding corn, which he prized as the greatest of all earthly possessions. And well might he prize it! He had no machine for converting corn into meal, but two stones, such as were used in the remote parts of the highlands of Scotland some years ago. And to beat the grain into meal by these two stones (a machine, remember, however imperfect) would occupy the labour of one-fourth of his family, to procure subsistence for the other three-fourths. The ancient Greeks, three thousand years ago, had improved upon the machinery of the hand-stones, for they had hand-mills. But Homer, the old Greek poet, describes the unhappy condition of the slave who was always employed in using this mill. The groans of the slave were unheeded by those who consumed the produce of his labour; and such was the necessity for meal, that the women were compelled to turn these mills when there were not slaves enough taken in war to perform this irksome office. There was plenty of labour then to be performed, even with the machinery of the hand-mill; but the slaves and the women did not consider that labour was a good in itself, and therefore they bitterly groaned under it. By and bye the understandings of men found out that water and wind would do the same work that the slaves and the women had done; and that a large quantity of labour was at liberty to be employed for other purposes. Does any one ask if society was in a worse state in consequence? We answer, labour is worth nothing without results. Its value is only to be measured by what it produces. If, in a country where hand-mills could be had, the people were to go on beating grain between two stones, all would pronounce them fools, because they could obtain an equal quantity of meal with a much less expenditure of labour. Some have a general prejudice against that sort of machinery which does its work with very little human assistance; it is not quite so certain, therefore, that they would agree that a people would be equal fools to use the hand-mill when they could employ the wind-mill or the water-mill. But we believe they would think that, if flour could drop from heaven, or be had like water by whoever chose to seek it, it would be the height of folly to have stones, or hand-mills, or water-mills, or wind-mills, or any machine whatever for manufacturing flour. Does any one ever think of *manufacturing* water? The cost of water is only the cost of the labour which brings it to the place in which it is consumed. Yet this admission overturns all objections against machinery. *We admit that it is desirable to obtain a thing with no labour at all; can we therefore doubt that it is desirable to obtain it with the least possible labour?* The only difference between no labour and a little labour is the difference of the cost of production. And the only difference between little labour and much labour is precisely the same. In procuring anything that administers to his necessities, man makes an exchange of his labour for the thing produced, and the less he gives of his labour the better of course is his bargain.

To return to the hand-mill and the water-mill. An ordinary water-mill for grinding corn will grind about thirty-six sacks a day. To do the same work with a hand-mill would require 150 men. At two shillings a day the wages of these men would amount to 15*l*., which, reckoning six working days, is 90*l*. a week, or 4680*l*. a year. The rent and taxes of a mill would be about 150*l*. a year, or 10*s*. a working day. The cost of machinery would be certainly more for the hand-mills than the water-mill, therefore we will not take the cost of machinery into calculation. To produce, therefore, thirty-six sacks of flour by hand we should pay 15*l*.; by the water-mill we should pay 10*s*.: that is, we should pay thirty times as much by the one process as by the other. The actual saving is something about the price of the flour in the market; that is, the consumer, if the corn were ground by hand, would pay double what he pays now for flour ground at a mill.

But if the system of grinding corn by hand were a very recent system of society, and the introduction of so great a benefit as the water-mill had all at once displaced the hand-grinders, as the spinning machinery displaced the spinning-wheel, what must become, it is said, of the one hundred and fifty men who earned the 15*l*. a-day, of which sum the consumer has now got 14*l*. 10*s*. in his pocket? They must go to other work. And what is to set them to that work? The same 14*l*. 10*s*.; which, being saved in the price of flour, gives the poor man, as well as the rich man, more animal food and fuel; a greater quantity of clothes, and of a better quality; a better house than his hand-labouring ancestors used to have, much as his house might yet be improved; better furniture, and more of it; domestic utensils; and books. To produce these things there must be more labourers employed than before. The quantity of labour is, therefore, not diminished, while its productiveness is much increased. It is as if every man among us had become suddenly much stronger and more industrious. The machines labour for us, and are yet satisfied without either food or clothing. They increase all our comforts, and they consume none themselves. The hand-mills are not grinding, it is true: but the ships are sailing that bring us foreign produce; the looms are moving that give us more clothes; the potter, and glass-maker, and joiner, are each employed to add to our household goods; we are each of us elevated in the scale of society; and all these things happen because machinery has diminished the cost of production.

Mill at Guy's Cliff.

The water-mill is, however, a simple machine compared with some mills of modern times. We are familiar with the village-mill. As we walk by the side of some gentle stream, such as that which turns the mill at Guy's Cliff, in Warwickshire, we hear at a distance the murmur of water and the growl of wheels. We come upon the old mill, such as it has stood for a couple of centuries. No peasant quarrels with the mill. It is an object almost of his love; for he knows that it cheapens his food. It seems a part of the natural scenery amidst which he has been reared.

But let a more efficient machine for grinding corn be introduced, such as the Americans have at Pittsburgh, and the peasant might think that the working millers would be ruined. And yet the mill at Pittsburgh is making flour cheaper in England, by that competition which here forces onward improvement in mill-machinery; and by increasing the consumption of flour calls into action more superintending labour for its production. That particular American mill produces 500 barrels of flour per day, each containing 196lbs., and it employs forty managing persons. It produces cheap flour by saving labour in all its processes. It stands on the bank of a navigable river—a high building into which the corn for grinding must be removed from boats alongside. Is the grain necessary to produce these 500 barrels of flour per day, amounting to 98,000lbs., carried by man's labour to the topmost floor of that high mill? It is "raised by an elevator consisting of an endless band, to which are fixed a series of metal cans revolving in a long wooden trough, which is lowered through the respective hatchways into the boat, and is connected at its upper end with the building where its belt is driven. The lower end of the trough is open, and, as the endless band revolves, six or eight men shovel the grain into the ascending cans, which raise it so rapidly that 4000 bushels can be lifted and deposited in the mill in an hour."[16] The drudging and unskilled labourers who would have toiled in carrying up the grain are free to do some skilled labour, of which the amount required is constantly increasing; and the cost saved by the elevator goes towards the great universal fund, out of which more labour and better labour are to find the means of employment.

[15] See an article by Mr. Bishop, on 'Locomotion of Animals,' in 'English Cyclopædia.'

[16] Whitworth's Special Report.

CHAPTER XI.

> Present and former condition of the country—Progress of cultivation—Evil influence of feudalism—State of agriculture in the sixteenth century—Modern improvements—Prices of wheat—Increased breadth of land under cultivation—Average consumption of wheat—Implements of agriculture now in use—Number of agriculturists in Great Britain.

It is the remark of foreigners, as they travel from the sea-coast to London, that the country is a garden. It has taken nineteen centuries to make it such a garden. The marshes in which the legions of Julius Cæsar had to fight, up to their loins, with the Britons, to whom these swamps were habitual, are now drained. The dense woods in which the Druids worshipped are now cleared. Populous towns and cheerful villages offer themselves on every side. Wherever the eye reaches there is cultivation. Instead of a few scattered families painfully earning a subsistence by the chace, or by tilling the land without the knowledge and the instruments that science has given to the aid of manual labour,—that cultivation is carried on with a systematic routine that improves the fertility of a good season, and diminishes the evils of a bad. Instead of the country being divided amongst hostile tribes, who have little communication, the whole territory is covered with a network of roads, and canals, and navigable rivers, and railroads, through which means there is an universal market, and wherever there is demand there is instant supply. Rightly considered, there is no branch of production which has so largely benefited by the power of knowledge as that of agriculture. It was ages before the great physical changes were accomplished which we now behold on every side; and we are still in a state of progress towards the perfection of those results which an over-ruling Providence had in store for the human race, in the gradual manifestation of those discoveries which have already so changed our condition and the condition of the world.

The history of cultivation in Great Britain is full of instruction as regards the inefficiency of mere traditional practice and the slowness with which scientific improvement establishes its dominion. It is no part of our plan to follow out this history; but a few scattered facts may not be without their value.

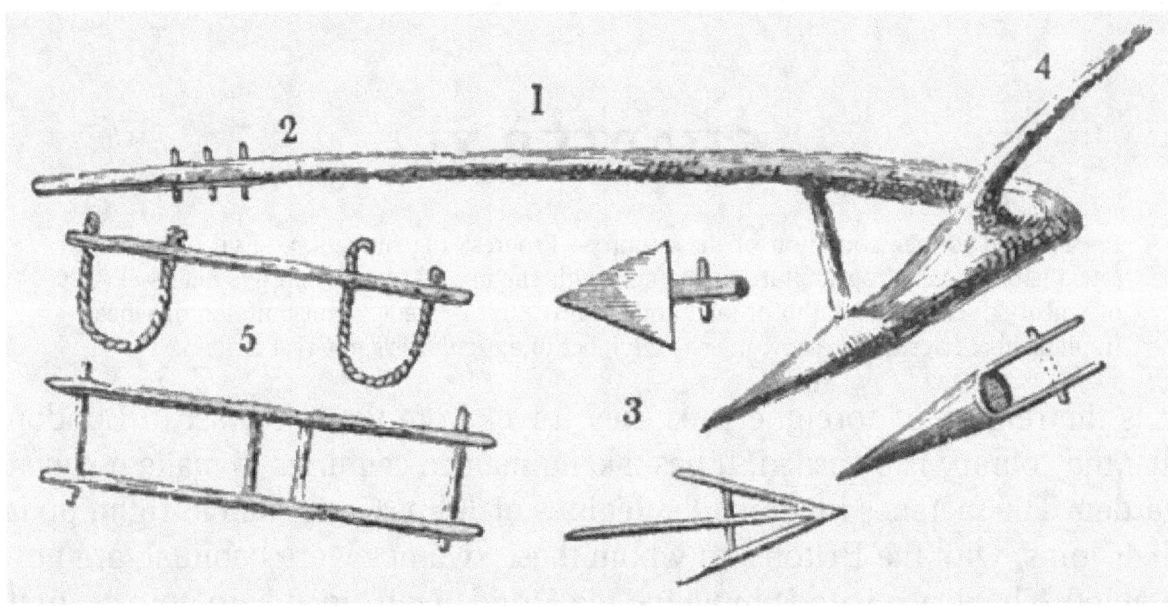

1. The plough. 2. The pole. 3. The share (various). 4. The handle, or plough-tail. 5. Yokes.

The oppressions of tenants that were perpetrated under the feudal system, when ignorant lords of manors impeded production by every species of extortion, may be estimated by one or two circumstances. There can be no doubt that the prosperity of a tenant is the best security for the landlord's due share of the produce of the land. Without manure, in some form or other, the land cannot be fertilized: the landlords knew this, but they required to have a monopoly of the fertility. Their tenants kept a few sheep, but the landlords reserved to themselves the exclusive privilege of having a sheepfold; so that the little tenants could not fold their own sheep on their own lands, but were obliged to let them be folded with those of their lord, or pay a fine.[17] The flour-mill was the exclusive property of the manorial lord, whether lay or ecclesiastical; and whatever the distance, or however bad the road, the tenant could grind nowhere but at the lord's mill. No doubt the rent of land was exceedingly low, and the lord was obliged to maintain himself and his dependents by adding something considerable to his means by many forms of legalized extortion. The rent of land was so low because the produce was inconsiderable, to an extent which will be scarcely comprehended by modern husbandmen. In the law-commentary called 'Fleta,' written about the end of the thirteenth century, the author says the farmer will be a loser unless corn be dear, if he obtains from an acre of wheat only three times the seed sown. He calculated the low produce at six bushels an acre: the average produce was perhaps little higher; we have distinct records of its being no higher a century afterwards. In 1390, at Hawsted, near Bury, the produce of the manor-farm was forty-two quarters of wheat, or three hundred and thirty-six bushels, from fifty-seven acres; and upon an average of three years sixty-one acres produced only seventy quarters, or five hundred and sixty bushels. Sir John Cullum, who collected these details from the records of his own property, says, "no particular dearness of corn followed, so that, probably, those very scanty crops were the usual and ordinary effects of the imperfect husbandry then practised." The husbandry was so imperfect that an unfavourable season for corn-crops, which in our days would have been compensated by a greater production of green crops, was followed by famine. When the ground was too hard, the seed could not be sown for want of the sufficient machine-power of plough and harrow. The chief instrument used was as weak and imperfect as the plough which we see depicted in Egyptian monuments, and which is still found in parts of Syria. The Oriental ploughman was with such an instrument obliged to bend over his plough, and load it with all the weight of his body, to prevent it merely scratching the ground instead of turning it up. His labour was great and his care incessant, as we may judge from the words of our Saviour,—"No man having put his hand to the plough, and looking back, is fit for the kingdom of God." Latimer, the Protestant martyr, in his 'Sermon of the Plough,' in which he holds that "preaching of the Gospel is one of God's plough-works, and the preacher is one of God's ploughmen," describes the labour upon which he raises his parallel: "For as the ploughman first setteth forth his plough, and then tilleth his land and breaketh it in furrows, and sometimes ridgeth it up again; and at another time harroweth it and clotteth it, and sometimes dungeth it and hedgeth it, diggeth it and weedeth it, purgeth it and maketh it clean,—so the prelate, the preacher, hath many divers offices to do." Latimer was the son of a Liecestershire farmer, and knew practically what he was talking about. He knew that the land would not bear an adequate crop without all this various and often-repeated labour. And yet the labour was so inadequately performed, that a few years after he had preached this famous sermon, we are told by a credible writer of the times of Queen Mary—William Bulleyn, a physician and botanist—that in 1555 "bread was so scant, insomuch that the plain poor people did make very much of acorns." A few years onward a great impulse was given to husbandry through various causes, amongst which the increased abundance of the precious metals through the opening of the mines of South America had no inconsiderable influence. The industrious spirit of England was fairly roused from a long sleep in the days of Queen Elizabeth. Harrison, in his 'Description of Britain,' says, "The soil is even now in these our days grown to be much more fruitful than it hath been in times past." This historian of manners saw the reason. "In times past" there was "idle and negligent occupation;" but when he wrote (1593) "our countrymen are grown to be more painful, skilful, and careful, through recompense of gain." The cultivators had their share of the benefits of cultivation; they had their "recompense of gain." Capital had been spread amongst the class of tenants: they were no longer miserable dependents upon their grasping lords. For a century or so onward the improvements in agriculture were not very decided. The rotation of crops was unknown; and winter food for sheep and cattle not being raised, the greater number were slaughtered and salted at Martinmas. The fleeces were wretchedly small, for the few sheep nibbled the stubbles and commons bare till the spring-time. The carcases of beef were not half their present size. At the beginning of the last century the turnip cultivation was introduced, and in the middle of the century the horse-hoeing husbandry came into use. Our agricultural revolution was fairly begun a hundred years ago; and yet for many years the value of manure was very imperfectly understood, and even up to our own time it has been wasted in every direction. There is given in Sir John Cullum's book an abstract of the lease of a farm in 1753. The tenant was to be allowed two shillings for every load of manure that he brought from Bury, about four miles distant. During twenty-one years the landlord was charged with only one load. At that period all agriculture was in a great degree traditional. There were no agricultural societies—no special journals for this great branch of national industry. Arthur Young applied his shrewd and observing talent to the dissemination of farming knowledge; but the agricultural mind, with very few exceptions, rejected book-knowledge as vain and impertinent. Chemistry applied to the soil was as unknown to the cultivator as the perturbations of the planets. Geology was an affair of conjecture, and had assumed no form of utility. Meteorology entered into no farmer's mode of estimating the comparative value of one site and another. Sir John Cullum made a most curious and instructive estimate of the science of the Suffolk farmers in 1784, when he wrote,—"The farm-houses are in general well furnished with every convenient accommodation. Into many of them a *barometer* has of late years been introduced—a most useful instrument for the husbandman, and which is mentioned here as *a striking instance of the intelligence of this period*."

The wars of the French Revolution, and the high prices consequent upon the almost utter absence of foreign supplies, gave a stimulus to agriculture, which principally displayed itself in the effort to bring waste lands into cultivation. From 1790 to 1819, a period of thirty years, there were two thousand one hundred and sixty-nine Inclosure Bills passed by Parliament. In the first ten years of this period the average price of wheat had increased ten shillings above the average of the ten years from 1780 to 1789. In the second ten years it had increased thirty-six shillings above that average. In the third ten years it was very nearly double, being 88*s*. 8*d*. from 1810 to 1819, and 45*s*. 9*d*. from 1780 to 1789. A portion of these prices, however, must be attributed to a depreciated currency. During that period of thirty years, very few of the great scientific improvements which have cheapened production had been introduced, although better modes of cultivation unquestionably prevailed. During the twenty years from 1820 to 1839 there were only three hundred and thirty-one Inclosure Bills passed; and the price of wheat had fallen to about the average of the ten years from 1790 to 1799, and it continued at that average for another ten years. In the ten years from 1840 to 1849, we cannot gather the amount of land brought into new cultivation from the number of Inclosure Bills, as there was a General Inclosure Act passed in 1835, to prevent the expense of particular bills for small tracts of land. But it has been calculated that, while more than three million acres were brought into cultivation in the twenty years from 1800 to 1819, about one million acres only were inclosed in the thirty years from 1820 to 1849.[18] If we look then, as we shall briefly do, at the wonderfully increased production of Great Britain, we shall be naturally led to the conclusion that some cause, much more efficient than the inclosure of waste lands, has given us the means of feeding a population which has doubled since the beginning of the century. This production is the result of the whole course of improvement effected by science, and stimulated by capital. The Bedford Level was drained by our ancestors. The fens of Cambridgeshire and Lincolnshire have been drained effectually in our time. That luxuriant flat now rejoices in waving corn-crops over many a mile. A few years ago it was a land of stagnant ditches; where little wind-mills, that looked to the solitary traveller through that cold district like the toys of children, lifted the water out of the trenches, and left an isolated acre or two of dry earth for man to toil in. Now mighty steam-pumps carry the water into artificial rivers, and the whole region is unrivalled for fertility.

It is estimated by some statists that the average consumption of wheat for each individual of the population is eight bushels. Others estimate that consumption at six bushels. It will be sufficient for a broad view of the increase of production, as compared with the increase of population, to take the consumption at eight bushels, or a quarter of wheat per head. In the ten years from 1801 to 1810, deducting an annual average of 600,000 quarters of foreign wheat and flour imported, the population in 1811 being 11,769,725, the number fed by wheat of home growth was somewhat above eleven millions. In the ten years from 1841 to 1850, deducting an annual average of 3,000,000 quarters of foreign wheat and flour, the population in 1851 being 21,121,967, the number fed by wheat of home growth was somewhat above eighteen millions. The productive power of the country had increased, in the course of fifty years, to the enormous extent of giving subsistence, in one article of agricultural produce alone, to seven millions of people. The population in 1751 was estimated at little more than seven millions. It has trebled in a century; and we may be perfectly sure that the production of the land has far more than trebled in that period. The probability is that it has quadrupled; for there is no doubt that the great bulk of the people are better fed than in 1751, when rye-bread was the common sustenance of the great body of labourers throughout the country.

Let us endeavour to take a rapid view of the implements of agriculture in common use at the present time—implements which have been described as "intended not to bring about new conditions of soil, nor to yield new products of any kind, but to do with more certainty and cheapness what had been done hitherto by employing the rude implements of former centuries." Such are the words of Mr. Pusey's admirable Report on the Agricultural Implements in the 'Exhibition of the Works of Industry of all Nations.' We cannot do better than furnish a few slight notices of the leading subjects of this report.

The plough and the harrow were the sole instruments of tillage at the beginning of this century. Bloomfield, in his 'Farmer's Boy,' describes them:

> "The ploughs move heavily, and strong the soil,
> And clogging harrows with augmented toil
> Dive deep."

The old plough used to be drawn with four horses; and they were needed. It was a cumbrous instrument, "adapted to the clay soils when those soils were the chief source of corn to the country, and had been handed down from father to son, after the heavy lands had been widely laid down to grazing-ground, and the former downs had become our principal arable land." The modern plough is an implement constructed on mathematical principles, which, by its mould-board, "raising each slice of earth (furrow-slice) from its flat position, through an upright one, lays it over half inclined on the preceding slice." The perfect instrument produces the skilled labourer. A good ploughman will set up a pole a quarter of a mile distant, and trace a furrow so true up to that goal that no eye can detect any divergence from absolute straightness. Mr. Pusey justly says that this is a triumph of art.

With an agriculture that permits no waste, much of the picturesque has fled from our fields. Bloomfield describes the repose of the ploughman after he has driven his team to the boundary of his furrow:—

> "Welcome green headland! firm beneath his feet;
> Welcome the friendly bank's refreshing seat;
> There, warm with toil, his panting horses browse
> Their sheltering canopy of pendent boughs."

Gone is the green headland; gone the cowslip bank; gone the pendent boughs. The furrow runs up to the extremest point of a vast field without hedges. Gone, too, are the green slips between the lands of common fields. Their very names of "balk" and "feather" are obsolete. These adornments of the landscape are inconsistent with the demands of a population that doubles itself in half a century. The labourer has small rest, and the soil has less. Under the old husbandry, before the culture of the green crops, one-third of the arable land was always idle. Two years of grain-crop, and one year of fallow, was the invariable rule. Look how the land is worked now.

The plough and the harrow turn up and pulverize the soil, but they do it much more effectually than of old. The roller is a noble iron instrument, instead of an old pollard. Modern ingenuity has added the clod-crusher. But something was still wanting for the better preparation of land for seed—this is the scarifier or cultivator; which, according to Mr. Pusey, will save one half of the horse-labour employed upon the plough. Into the details of this saving it is no part of our purpose to enter.[19] We give a cut of the implement, covering as much ground in width as 8-1/2 ploughs.

Clod-crusher.

Scarifier.

We proceed to "Instruments used in the Cultivation of Crops." Mr. Pusey tells us that "the sower with his seed-lip has almost vanished from southern England, driven out by a complicated machine, the drill, depositing the seed in rows, and drawn by several horses." We miss the sower; and the next generation may require a commentary upon the many religious and moral images that arose out of his primitive occupation. When James Montgomery says of the seed of knowledge, "broadcast it o'er the land," some may one day ask what "broadcast" means. But the drill does not only sow the seed; it deposits artificial manures for the reception of the seed. The bones that were thrown upon the dunghill are now crushed. The mountains of fertilizing matter that have been accumulated through ages on islands of the Pacific, from the deposits of birds resting in their flight upon rocks of that ocean, and which we call guano, now form a great article of commerce. Superphosphate, prepared from bones, or from the animal remains of geological ages, is another of the precious dusts which the drill economizes. There are even drills for dropping water combined with superphosphate. "Such," says Mr. Pusey, "is the elastic yet accurate pliability with which, in agriculture, mechanism has seconded chemistry." The system of horse-hoeing, which is the great principle of modern husbandry, entirely depends upon the use of the drill. The horse-hoe cannot be worked unless the plants are in rows. Such a hoe as this will clean at once nine rows of wheat, six of beans, and four of turnips. To manage such an instrument requires "a steady and cool hand." The skilled labourer is as essential as the beautiful machine.

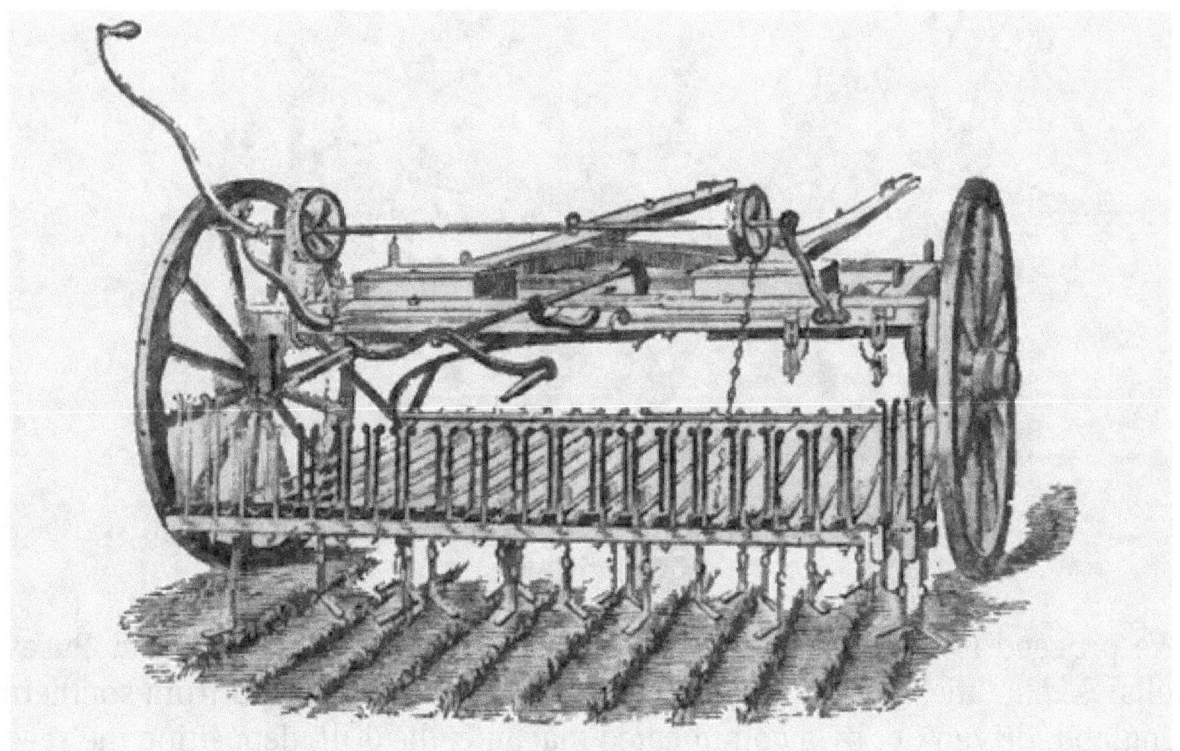

Horse-hoe

Of instruments for gathering the harvest, the most important are reaping-machines. In the United States they are sold to a great extent. Mr. M'Cormick, who completed his invention in 1845, states that the demand reaches to a thousand annually. Mr. Pusey says of this machine that, "in bad districts and late seasons, it may often enable the farmer to save the crop." In Scotland and the north of England Mr. Bell's reaping-machine is coming into extensive use. The Americans have also their mowing-machines, drawn by two horses, which mow, upon an average, six acres of grass per day. The haymaking machines, as labour-saving instruments, are not uncommon in England.

Machines for preparing corn for market are amongst the most important inventions of modern times. Here, indeed, agriculture assumes many of the external features of a manufacture. Steam comes prominently into action. In many large farms there is fixed steam-power; and most efficient it is. But the moveable steam-engine comes to the aid of the small farmer; and in some districts that power is let out to those who want it. By this little engine applied to a thrashing-machine, corn is thrashed at once from the rick, instead of being carried into the barn. Here is a representation of the combined steam-engine and thrashing-machine. The thrashing-machine with horse-power is that generally used in England. Rarely, now, can the beautiful description of Cowper be realized:—

"Thump after thump resounds the constant flail,
That seems to swing uncertain, and yet falls
Full on the destined ear."

Moveable steam-engine and thrashing-machine.

Few now wield that ancient instrument. Nor is the chaff now separated from the corn by the action of the wind, which was called winnowing, but we have the winnowing-machine, by which forty quarters of wheat can be taken from the thrashing-machine and prepared for the market in five hours.

But machinery does not end here. The food of stock is prepared by machines. First, there is the turnip-cutter. Our 'Farmer's Boy' will tell us how his sheep and kine were fed in the winter fifty years ago:—

Thrashing-machine with horse-power.

"No tender ewe can break her nightly fast,
Nor heifer strong begin the cold repast,
Till Giles with ponderous beetle foremost go,
And scattering splinters fly at every blow;
When, pressing round him, eager for the prize,
From their mix'd breath warm exhalations rise."

We are told that "lambs fed with a turnip-cutter would be worth more at the end of a winter by eight shillings a head than lambs fed on whole turnips." The chaff-cutter is an instrument equally valuable.

The last machine which we shall mention is connected with the greatest of all improvements in the crop-producing power of British land—the system of tile-draining. Pipes are now made by machinery; and land may be effectually drained at a cost of 4*l.* per acre.

Draining-tile machine.

The farmers of England have made what we may fairly call heroic efforts to meet foreign competition; but their efforts would have been comparatively vain had science not come to the aid of production.

According to the Census of 1851, the total population of Great Britain is 20,959,477—in round numbers, twenty-one millions. In the 'Return of Occupations,' one-half of this entire population is found under the family designation—such as child at home, child at school, wife, daughter, sister, niece, with no particular occupation attributed to them. They are important members of the state; they are growing into future producers, or they preside over the household comforts, without which there is little systematic industry. But they are not direct producers. Of this half of the entire population, one-fifth belong to the class of cultivators, viz.:—

	Male.	Female.
Holders of farms	275,676	28,044

Farmers' relatives, in-door	137,446	
Out-door labourers	1,006,728	70,899
Farm-servants, in-door	235,943	128,251
Shepherds, out-door	19,075	
Woodmen	9,832	
Gardeners	78,462	2,484
Farm bailiffs	12,805	
Graziers	3,036	
	1,779,003	229,678

This total (in which we omit the farmers' wives and daughters, amounting to about 240,000) shows that one-fifth of the working population provide food, with the exception of foreign produce, for themselves and families and the other four-fifths of the population. Such a result could not be accomplished without the appliances of scientific power which we have described in this chapter. In the early steps of British society a very small proportion of labour could be spared for other purposes than the cultivation of the soil. It has been held that a community is considerably advanced when it can spare one man in three from working upon the land. Only twenty-six per cent of our adult males are agricultural—that is, three men labour at some other employment, while one cultivates the land. During the last forty years the proportion of agricultural employment, in comparison with manufacturing and commercial, has been constantly decreasing; and is now about twenty per cent., whereas in 1811 it was thirty-five per cent. of all occupations.

[17] Cullum's 'History of Hawsted.'

[18] See various tables in Porter's 'Progress of the Nation.'

[19] See 'Journal of Royal Agricultural Society,' vol. xii. p. 595.

CHAPTER XII.

Production of a knife—Manufacture of iron—Raising coal—The hot-blast—Iron bridges—Rolling bar-iron—Making steel—Sheffield manufactures—Mining in Great Britain—Numbers engaged in mines and metal manufactures.

We have been speaking somewhat fully of agricultural instruments and agricultural labour, because they are at the root of all other profitable industry. Bread and beef make the bone and sinew of the workman. Ploughs and harrows and drills and thrashing-machines are combinations of wood and iron. Rude nations have wooden ploughs. Unless the English labourer made a plough out of two pieces of stick, and carried it upon his shoulder to the field, as the toil-worn and poor people of India do, he must have some iron about it. He cannot get iron without machinery. He cannot get even his knife, his tool of all-work, without machinery. From the first step to the last in the production of a knife, machinery and scientific appliances have done the chief work. People that have no science and no machinery sharpen a stone, or bit of shell or bone, and cut or saw with it in the best way they can; and after they have become very clever, they fasten it to a wooden handle with a cord of bark. An Englishman examines two or three dozens of knives, selects which he thinks the best, and pays a shilling for it, the seller thanking him for his custom. The man who has nothing but the bone or the shell would gladly toil a month for that which does not cost an English labourer half a day's wages.

And how does the Englishman obtain his knife upon such easy terms? From the very same cause that he obtains all his other accommodations cheaper, in comparison with the ordinary wages of labour, than the inhabitant of most other countries—that is, from the operations of science, either in the making of the thing itself, or in procuring that without which it could not be made. We must always remember that, if we could not get the materials without scientific application, it would be impossible for us to get what is made of those materials—even if we had the power of fashioning those materials by the rudest labour.

Keeping this in mind, let us see how a knife could be obtained by a man who had nothing to depend upon but his hands.

Ready-made, without the labour of some other man, a knife does not exist; but the iron, of which the knife is made, is to be had. Very little iron has ever been found in a native state, or fit for the blacksmith. The little that has been found in that state has been found only very lately; and if human art had not been able to procure any in addition to that, gold would have been cheap as compared with iron.

Iron is, no doubt, very abundant in nature; but it is always mixed with some other substance that not only renders it unfit for use, but hides its qualities. It is found in the state of what is called *iron-stone*, or *iron-ore*. Sometimes it is mixed with clay, at other times with lime or with the earth of flint; and there are also cases in which it is mixed with sulphur. In short, in the state in which iron is frequently met with, it is a much more likely substance to be chosen for paving a road, or building a wall, than for making a knife.

But suppose that the man knows the particular ore or stone that contains the iron, how is he to get it out? Mere force will not do, for the iron and the clay, or other substance, are so nicely mixed, that, though the ore were ground to the finest powder, the grinder is no nearer the iron than when he had a lump of a ton weight.

A man who has a block of wood has a wooden bowl in the heart of it; and he can get it out too by labour. The knife will do it for him in time; and if he take it to the turner, the turner with his machinery, his lathe, and his gouge, will work it out for him in half an hour. The man who has a lump of iron-ore has just as certainly a knife in the heart of it; but no mere labour can work it out. Shape it as he may, it is not a knife, or steel, or even iron—it is iron-ore; and dress it as he will, it would not cut better than a brickbat—certainly not so well as the shell or bone of the savage.

There must be knowledge before anything can be done in this case. We must know what is mixed with the iron, and how to separate it. We cannot do it by mere labour, as we can chip away the wood and get out the bowl; and therefore we have recourse to fire.

In the ordinary mode of using it, fire would make matters worse. If we put the material into the fire as a stone, we should probably receive it back as slag or dross. We must, therefore, prepare our fuel. Our fire must be hot, very hot; but if our fuel be wood we must burn it into charcoal, or if it be coal into coke.

The charcoal, or coke, answers for one purpose; but we have still the clay or other earth mixed with our iron, and how are we to get rid of that? Pure clay, or pure lime, or pure earth of flint, remains stubborn in our hottest fires; but when they are mixed in a proper proportion, the one melts the other.

So charcoal or coke, and iron-stone or iron-ore, and limestone, are put into a furnace; the charcoal or coke is lighted at the bottom, and wind is blown into the furnace, at the bottom also. If that wind is not sent in by machinery, and very powerful machinery too, the effect will be little, and the work of the man great; but still it can be done.

In this furnace the lime and clay, or earth of flint, unite, and form a sort of glass, which floats upon the surface. At the same time the carbon, or pure charcoal, of the fuel, with the assistance of the limestone, mixes with the stone, or ore, and melts the iron, which, being heavier than the other matters, runs down to the bottom of the furnace, and remains there till the workman lets it out by a hole made at the bottom of the furnace for that purpose, and plugged with sand. When the workman knows there is enough melted, or when the appointed time arrives, he displaces the plug of sand with an iron rod, and the melted iron runs out like water, and is conveyed into furrows made in sand, where it cools, and the pieces formed in the principal furrows are called "sows," and those in the furrows branching from them "pigs."

We are now advanced a considerable way towards the production of a knife. We have the materials of a knife. We have the iron extracted out of the iron-ore. Before we trace the progress of a knife to its final polish, let us see what stupendous efforts of machinery have been required to produce the cast iron.

In every part of the operation of making iron—in smelting the iron out of the ore; in moulding cast iron into those articles for which it is best adapted;

in working malleable iron, and in applying it to use after it is made; nothing can be done without fire, and the fuel that is used in almost every stage of the business is coal. The coal trade and the iron trade are thus so intimately connected, so very much dependent upon each other, that neither of them could be carried on to any extent without the other. The coal-mines supply fuel, and the iron-works give mining tools, pumps, railroads, wheels, and steam-engines, in return. A little coal might be got without the iron engines, and a little iron might be made without coals, by the charcoal of wood. But the quantity of both would be trifling in comparison. The wonderful amount of the production of iron in Great Britain, and the cheapness of iron, as compared with the extent of capital required for its manufacture, arises from the fact that the coal-beds and the beds of iron-ore lie in juxta-position. The iron-stones alternate with the beds of coal in almost all our coal-fields; and thus the same mining undertakings furnish the ore out of which iron is made and the fuel by which it is smelted. If the coal were in the north, and the fuel in the south, the carriage of the one to the other would double the cost.

There was a time when iron was made in this country with very little machinery. Iron was manufactured here in the time of the Romans; but it was made with great manual labour, and was consequently very dear. Hutton, in his 'History of Birmingham,' tells us that there is a large heap of cinders near that town which have been produced by an ancient iron-furnace; and that from the quantity of cinders, as compared with the mechanical powers possessed by our forefathers, the furnace must have been constantly at work from the time of Julius Cæsar. A furnace with a steam blast would produce as large a heap in a few years.

At present a cottager in the south of England, where there is no coal in the earth, may have a bushel of good coals delivered at the door of his cottage for eighteen pence; although that is far more than the price of coal at the pit's mouth. If he had even the means of transporting himself and his family to the coal district, he could not, without machinery, get a bushel of coals at the price of a year's work. Let us see how a resolute man would proceed in such an undertaking.

The machinery, we will say, is gone. The mines are filled up, which the greater part of them would be, with water, if the machinery were to stop a single week. Let us suppose that the adventurous labourer knows exactly the spot where the coal is to be found. This knowledge, in a country that has never been searched for coals before, is no easy matter, even to those who understand the subject best: it is the province of geology to give that knowledge. But we shall suppose that he gets over that difficulty too, for after it there is plenty of difficulty before him.

Well, he comes to the exact spot that he seeks, and places himself right over the seam of coal. That seam is only a hundred fathoms below the surface, which depth he will, of course, reach in good time. To work he goes; pares off the green sod with his shovel, loosens the earth with his pickaxe, and, in the course of a week, is twenty feet down into the loose earth and gravel, and clears the rock at the bottom. He rests during the Sunday, and comes refreshed to his work on Monday morning; when, behold, there are twelve feet of water in his pit.

Suppose he now calls in the aid of a bucket and rope, and that he bales away, till, as night closes, he has lowered the water three feet. Next morning it is up a foot and a half: but no matter; he has done something, and next day he redoubles his efforts, and brings the water down to only four feet. That is encouraging; but, from the depth, he now works his bucket with more difficulty, and it is again a week before his pit is dry. The weather changes; the rain comes down heavily; the surface on which it falls is spongy; the rock which he has reached is water-tight; and in twelve hours his pit is filled to the brim. It is in vain to go on.

The sinking of a pit, even to a less depth than a hundred fathoms, sometimes demands, notwithstanding all the improvements by machinery, a sum of not less than a hundred pounds a fathom, or ten thousand pounds for the whole pit; and therefore, supposing it possible for a single man to do it at the rate of eighteen pence a day, the time which he would require would be between four hundred and five hundred years.

Whence comes it that the labour of between four hundred and five hundred years is reduced to a single day? and that which, independently of the carriage, would have cost ten thousand pounds, is got for eighteen pence? It is because man joins with man, and machinery is employed to do the drudgery. Nations that have no machinery have no coal fires, and are ignorant that there is hidden under the earth a substance which contributes more, perhaps, to the health and comfort of the inhabitants of Britain than any other commodity which they enjoy.

No nations have worked coal to anything approaching the extent in which it has been worked by our countrymen. It has been calculated that France, Belgium, Spain, Prussia, Bohemia, and the United States of America, do not annually produce more than seventeen million tons of coal, which is about half of our annual produce.[20]

Steam-Boiler making.

The greater part of the coal now raised in Britain is produced by the employment of the most enormous mechanical power. There are in some places shallow and narrow pits, where coals may be raised to the surface by a windlass; and there are others where horse-power is employed. But the number of men that can work at a windlass, or the number of horses that can be yoked to a gin, is limited. The power of the steam-engine is limited only by the strength of the materials of which it is formed. The power of a hundred horses, or of five hundred men, may be very easily made by the steam-engine to act constantly, and on a single point; and thus there is scarcely anything in the way of mere force which the engine cannot be made to do. We have seen a pit in Staffordshire, which hardly gave coal enough to maintain a cottager and his family, for he worked the pit with imperfect machinery—with a half-starved ass applied to a windlass. A mile off was a steam-engine of 200-horse power, raising tons of coals and pumping out rivers of water with a force equal to at least a thousand men. This vast force acted upon a point; and therefore no advantage was gained over the machine by the opposing force of water, or the weight of the material to be raised. Before the steam-engine was invented, the produce of the coal-mines barely paid the expense of working and keeping them dry; and had it not been for the steam-engines and other machinery, the supply would long before now have dwindled into a very small quantity, and the price would have become ten or twenty times its present amount. The quantity of coal raised in Great Britain was estimated by Professor Ansted in 1851 at thirty-five million tons; and the value at nine millions sterling at the pit-mouth, and eighteen millions at the place of consumption. The capital engaged in the coal trade was then valued at ten millions. In 1847 the annual value of all the precious metals raised throughout the world was estimated at thirteen millions sterling. That value has been increased within a few years. But the coal of Great Britain, as estimated by the cost at the pit's mouth, is above two-thirds of this value of the precious metals seven years ago; and the mean annual value, at the furnace, of iron smelted by British coal being eight millions sterling, the value together of our iron and our coal exceeds the value of all the gold and silver of South America, and California, and Australia, however large that amount has now become.

How the value of our cast iron has been increased by modern science may be in some degree estimated by a consideration of what the hot-blast has accomplished. The hot-blast blows hot air into the iron-furnace instead of cold air. The notion seems simple, but the results are wonderful. The inventor, Mr. Neilson, has seen since 1827 the production of iron raised from less than seven hundred thousand tons to two million two hundred thousand tons. The iron is greatly cheaper than a quarter of a century ago, for only about one-half the coal formerly used is necessary for its production. That production is almost unlimited in amount. In 1788 we produced only sixty thousand tons, or one-thirty-sixth part of what we now produce. The beautiful iron bridge of Colebrook-dale, erected in 1779, consumed three hundred and seventy-eight tons of cast iron. The wonderful Britannia Bridge which has been carried over the Menai Strait, hung in mid air at the height of a hundred feet above the stream, has required ten thousand tons of iron for its completion. If chemistry and machinery had not been at work to produce more iron and cheaper iron, how would our great modern improvements have stopped short—our railroads, our water-pipes, our gas-pipes, our steam-ships! How should we have lacked the great material of every useful implement, from the gigantic anchor that holds the man-of-war firm in her moorings, and the mighty gun that, in the last resort, asserts a spirit without which all material improvement cannot avert a nation's decay,—to the steel pen with which thoughts are exchanged between friends at the opposite ends of the earth, and the needle by which the poor seamstress in her garret maintains her place amongst competing numbers.

The first iron bridge, Colebrook Dale.

Nearly all the people now engaged in iron-works are supported by the improvements that have been made in the manufacture, *by machinery*, since 1788. Yes, wholly by the machinery; for before then the quantity made by the charcoal of wood had fallen off one-fourth in forty-five years. The wood for charcoal was becoming exhausted, and nothing but the powerful blast of a machine will make iron with coke. Without the aid of machinery the trade would have become extinct. The iron and the coal employed in making it would have remained useless in the mines.

And now, having seen what is required to produce a "pig" of cast iron, let us return to the knife, whose course of manufacture we traced a little way.

The lump of cast iron as it leaves the furnace has many processes to go through before it becomes fit for making a knife. It cannot be worked by the hammer, or sharpened to a cutting edge; and so it must be made into malleable iron,—into a kind of iron which, instead of melting in the fire, will soften, and admit of being hammered into shape, or united by the process of welding.

The methods by which this is accomplished vary; but they in general consist in keeping the iron melted in a furnace, and stirring it with an iron rake, till the blast of air in the furnace burns the greater part of the carbon out of it. By this means it becomes tough; and, without cooling, is taken from the furnace and repeatedly beaten by large hammers, or squeezed through large rollers, until it becomes the bar-iron of which so much use is made in every art of life.

Rolling bar-iron.

About the close of the last century the great improvement in the manufacture of bar-iron was introduced by passing it through grooved rollers, instead of hammering it on the anvil; but in our own time the invention has become most important. The inventor, Mr. Coet, spent a fortune on the enterprise and died poor. His son, in 1812, petitioned Parliament to assign him some reward for the great gift that his father had bestowed upon the nation. He asked in vain. It is the common fate of the ingenious and the learned; and it is well that life has some other consolations for the man that has exercised his intellect more profitably for the world than for himself, than the pride of the mere capitalist, who thinks accumulation, and accumulation only, the chief business of existence. Rolling bar-iron is one of the great labour-saving principles that especially prevail in every branch of manufacture in metals. The unaided strength of all the men in Britain could not make all the iron which is at present made, though they did nothing else. Machinery is therefore resorted to; and water-wheels, steam-engines, and all sorts of powers are set to work in moving hammers, turning rollers, and drawing rods and wires through holes, till every workman can have the particular form which he wants. If it were not for the machinery that is employed in the manufacture, no man could obtain a spade for less than the price of a year's labour; the yokes of a horse would cost more than the horse himself; and the farmer would have to return to wooden plough-shares, and hoes made of sticks with crooked ends.

After all this, the iron is not yet fit for a knife, at least for such a knife as an Englishman may buy for a shilling. Many nations would, however, be thankful for a little bit of it, and nations too in whose countries there is no want of iron-ore. But they have no knowledge of the method of making iron, and have no furnaces or machinery. When our ships sail among the people of the eastern islands, those people do not ask for gold. "Iron, iron!" is the call; and he who can exchange his best commodity for a rusty nail or a bit of iron hoop is a fortunate individual.

We are not satisfied with that in the best form, which is a treasure to those people in the worst. We must have a knife, not of iron, but of *steel*,—a substance that will bear a keen edge without either breaking or bending. In order to get that, we must again change the nature of our material.

How is that to be done? The oftener that iron is heated and hammered, it becomes the softer and more ductile; and as the heating and hammering forced the carbon out of it, if we give it the carbon back again, we shall harden it; but it happens that we also give it other properties, by restoring its carbon, when the iron has once been in a ductile state.

For this purpose, bars or pieces of iron are buried in powdered charcoal, covered up in a vessel, and kept at a red heat for a greater or less number of hours, according to the object desired. There are niceties in the process, which it is not necessary to explain, that produce the peculiar quality of steel, as distinguished from cast iron. If the operation of heating the iron in charcoal is continued too long, or the heat is too great, the iron becomes cast steel, and cannot be welded; but if it is not melted in the operation, it can be worked with the hammer in the same manner as iron.

In each case, however, it has acquired the property upon which the keenness of the knife depends; and the chief difference between the cast steel, and the steel that can bear to be hammered is, that cast steel takes a keener edge, but is more easily broken.

Shear and Tilt Hammers: Steel-manufacture.

The property which it has acquired is that of bearing to be tempered. If it be made very hot, and plunged into cold water, and kept there till it is quite cooled, it is so hard that it will cut iron, but it is brittle. In this state the workman brightens the surface, and lays the steel upon a piece of hot iron, and holds it to the fire till it becomes of a colour which he knows from experience is a test of the proper state of the process. Then he plunges it again into water, and it has the degree of hardness that he wants.

The grinding a knife, and the polishing it, even when it has acquired the requisite properties of steel, if they were not done by machinery, would cost more than the whole price of a knife upon which machinery is used. A travelling knife-grinder, with his treadle and wheels, has a machine, but not a very perfect one. The Sheffield knife-maker grinds the knife at first upon wheels of immense size, turned by water or steam, and moving so quickly that they appear to stand still—the eye cannot follow the motion. With these aids the original grinding and polishing cost scarcely anything; while the travelling knife-grinder charges two pence for the labour of himself and his wheel in just sharpening it.

File-cutters.

The "Sheffield whittle" is as old as the time of Edward III., as we know from the poet Chaucer. Sheffield is still the metropolis of steel. It is in the change of iron into steel by a due admixture of carbon—by hammering, by casting, by melting—that the natural powers of Sheffield, her water and her coal, have become of such value. Wherever there is a stream with a fall, there is the grinding-wheel at work: and in hundreds of workshops the nicer labour of the artificer is fashioning the steel into every instrument which the art of man can devise, from the scythe of the mower to the lancet of the surgeon. The machinery that made the steel has called into action the skill that makes the file-cutter. No machine can make a file. The file-cutter with a small hammer can cut notch after notch in a piece of softened steel, without a guide or gauge,—even to the number of a hundred notches in an inch. It is one out of many things in which skilled labour triumphs over the uniformity of operation which belongs to a machine. The cutting of files alone in Great Britain gives employment to more than six thousand persons. This is one of the many instances in which it is evident that the application of machinery to the arts calls into action an almost infinite variety of handicrafts. An ordinary workman can obtain a knife for the price of a few hours' labour. The causes are easily seen. Every part of the labour that can be done by machinery is so done. One turn of a wheel, one stroke of a steam-engine, one pinch of a pair of rollers, or one blow of a die, will do more in a second than a man could do in a month. One man, also, has but one thing to do in connexion with the machinery; and when the work of the hand succeeds to the work of the wheel or the roller, the one man, like the file-cutter, has still but one thing to do. In course of time he comes to do twenty times as much as if he were constantly shifting from one thing to another. The value of the work that a man does is not to be measured in all cases by the time and trouble that it cost him individually, but by the market value of what he produces; which value is determined, as far as labour is concerned, by the price paid for doing it in the best and most expeditious mode.

And does not all this machinery, and this economy of labour, it may still be said, deprive many workmen of employment? No. By these means the iron trade gives bread to hundreds, where otherwise it would not have given bread to one. There are more hands employed at the iron-works than there would have been if there had been no machinery; because without machinery men could not produce iron cheap enough to be generally used.

The machinery that is now employed in the iron trade, not only enables the people to be supplied cheaply with all sorts of articles of iron, but it enables a great number of people to find employment, not in the iron trade only, but in all other trades, who otherwise could not have been employed; and it enables everybody to do more work with the same exertion by giving them better tools; while it makes all more comfortable by furnishing them with more commodious domestic utensils.

There are thousands of families on the face of the earth, that would be glad to exchange all they have for a tin kettle, or an iron pot, which can be bought anywhere in the three kingdoms for a shilling or two. And could the poor man in this country but once see how even the rich man in some other places must toil day after day before he can scrape or grind a stone so as to be able to boil a little water in it, or make it serve for a lamp, he would account himself a poor man no more. An English gipsy carries about with him more of the conveniences of life than are enjoyed by the chiefs or rulers in countries which naturally have much finer climates than that of England. But they have no machinery, and therefore they are wretched.

Great Britain is a country rich in other minerals than iron-stone and coal. Our earliest ancestors are recorded to have exchanged tin with maritime people who came to our shores. They had lead also, which was cast into oblong blocks during the Roman occupation of the island, and which bear the imperial stamp. At the beginning of the eighteenth century we worked tin into pewter, which, in the shape of plates, had superseded wooden trenchers. But we raised and smelted no copper, importing it unwrought. The valuable tin and copper mines of Cornwall were imperfectly worked in the middle of the last century, because the water which overflowed them was only removed by hydraulic engines, the best of which was introduced in 1700. When Watt had reconstructed the steam-engine, steam-power began to be employed in draining the Cornwall mines. In 1780, 24,443 tons of copper-ore were raised, producing 2932 tons of copper. In 1850, 155,025 tons of ore were obtained, producing 12,254 tons of copper. The tin-mines produced 1600 tons in 1750, and 10,719 tons in 1849. The produce of the lead-mines has not been accurately estimated.

Entrance to the Mine of Odin, an ancient Lead-mine in Derbyshire.

In all mining operations, conducted as they are in modern times, and in our own country, we must either go without the article produced, whether coal, or iron, or lead, or copper, if the machines were abolished,—or we must employ human labour, in works the most painful, at a price which would not only render existence unbearable, but destroy it altogether. The people, in that case, would be in the condition of the unhappy natives of South America, when the Spaniards resolved to get gold at any cost of human suffering. The Spaniards had no machines but pickaxes and spades to put in the hands of the poor Indians. They compelled them to labour incessantly with these, and half the people were destroyed. Without machinery, in places where people can obtain even valuable ore for nothing, the collection and preparation of metals is hardly worth the labour. Mungo Park describes the sad condition of the Africans who are always washing gold-dust;—and we have seen in Derbyshire a poor man separating small particles of lead from the limestone, or spar, of that country, and unable to earn a shilling a day by the process. A man of capital erects lead-works, and in a year or two obtains an adequate profit, and employs many labourers.

It may enable us, in addition to our slight notices of quantities produced, to form something like an accurate conception of the vast mineral industry of this country, if we give the aggregate of men employed as miners and metal-workers, according to the census of 1851. Of coal-miners there were 216,366; of iron-miners, 27,098; of copper-miners, 18,468; of tin-miners, 12,912; of lead-miners, 21,617. This is a total of 296,461. In the manufacture of various articles of iron and steel, in addition to the iron and coal miners, who cannot be accurately distinguished, there are employed 281,578 male workers, and 18,807 female; and in the manufacture of articles of brass and other mixed metals, 46,076; of which number 8370 are females. The workers in metal thus enumerated amount to 542,922. We may add, from the class of persons engaged in mechanic productions, in which we find 48,050 engine and machine makers, and 7429 gunsmiths, a number that will raise the aggregate of miners and workers in metals to 600,000 persons. The boldness of some of the operations which are conducted in this department of industry, the various skill of the labourers, and the vastness of the aggregate results, impress the mind with a sense of power that almost belongs to the sublime. The fables of mythology are tame when compared with these realities of science. Vulcan, with his anvils in Ætna, is a feeble instrument by the side of the steam-hammer that forges an anchor, or the hydraulic press that lifts a bridge. A knot of Cupids co-operating for the fabrication of their barbed arrows is the poetry of painting applied to the arts. But there is higher poetry in that triumph of knowledge, and skill, and union of forces, which fills a furnace with fifty thousand pounds of molten iron, and conducts the red-hot stream to the enormous mould which is to produce a cylinder without a flaw.

CHAPTER XIII.

>Conveyance and extended use of coal—Consumption at various periods—Condition of the roads in the seventeenth and eighteenth centuries—Advantages of good roads—Want of roads in Australia—Turnpike-roads—Canals—Railway of 1680—Railway statistics.

We have seen how by machinery more than thirty-five million tons of coal—now become one of the very first necessaries of life—are obtained, which without machinery could not be obtained at all in the thousandth part of the quantity; and which, consequently, would be a thousand times the price—would, in fact, be precious stones, instead of common fuel.

Engines or machines, of some kind or other, not only keep the pits dry and raise the coals to the surface, but convey them to the ship upon railroads; the ship, itself a machine, carries them round all parts of the coast; barges and boats convey them along the rivers and canals; and, within these few years, railways have carried the coals of the north into remote places in the southern and other counties, where what was called "sea-coal," from its being carried coastwise, was scarcely known as an article of domestic use. The inhabitants of such places had no choice but to consume wood and turf for every domestic purpose.

Through the general consumption of wood instead of coal, a fire for domestic use in France is a great deal dearer than a fire in England; because, although the coal-pits are not to be found at every man's door, nor within many miles of the doors of some men, machinery at the pits, and ships and barges, and railways, which are also machinery, enable most men to enjoy the blessings of a coal fire at a much cheaper rate than a fire of wood, which is not limited in its growth to any particular district. Without the machinery to bring coals to his door, not one man out of fifty of the present population of England could have had the power of warming himself in winter; any more than without the machines and implements of farming he could obtain food, or without those of the arts he could procure clothing. The sufferings produced by a want of fuel cannot be estimated by those who have abundance. In Normandy, very recently, such was the scarcity of wood, that persons engaged in various works of hand, as lacemaking by the pillow, absolutely sat up through the winter nights in the barns of the farmers, where cattle were littered down, that they might be kept warm by the animal heat around them. They slept in the day, and were warmed by being in the same outhouse with cows and horses at night;—and thus they worked under every disadvantage, because fuel was scarce and very dear.

Coals were consumed in London in the time of Queen Elizabeth; but their use was, no doubt, very limited. Shakspere, who always refers to the customs of his own time, makes Dame Quickly speak of "sitting in my Dolphin-chamber at the round table, by a sea-coal fire, on Wednesday in Whitsun week." But Mrs. Quickly was a luxurious person, who had plate and tapestry and gilt goblets. Harrison, in his 'Description of Britain,' at the same period, says, that coal is "used in the cities and towns that lie about the coast;" but he adds, "I marvel not a little that there is no trade of these into Sussex and Southamptonshire; for want thereof the smiths do work their iron with charcoal." He adds, with great truth, "I think that far carriage be the only cause."

The consumption of coal in London in the last year of Charles II. (1685) amounted to three hundred and fifty thousand tons. This was really a large consumption, however insignificant it may sound when compared with the modern demand of the metropolis. In 1801 there were imported into London about a million tons of coals. In 1850, three million six hundred thousand tons were brought to the London market. The average contract price in the ten years ending 1810 was 45*s*. 6*d*.; in the ten years ending 1850 it was 18*s*. 6*d*. But in 1824 the oppressive duty of 7*s*. 6*d*. per ton on seaborne coals was reduced to 4*s*.; and in 1831 the duty was wholly repealed. It is the boast of our present fiscal system that the chief materials of manufacture, and the great necessaries and conveniences of life, are no longer made dear by injudicious taxation.

The chief power which produces coal and iron cheap is that of machinery. It is the same power which distributes these bulky articles through the country, and equalizes the cost in a considerable degree to the man who lives in London and the man who lives in Durham or Staffordshire. The difference in cost is the price of transport; and machinery, applied in various improved ways, is every year lessening the cost of conveyance, and thus equalizing prices throughout the British Islands. The same applications of mechanical power enable a man to move from one place to another with equal ease, cheapness, and rapidity. Quick travelling has become cheaper than slow travelling. The time saved remains for profitable labour.

About a hundred and ninety years ago, when the first turnpike-road was formed in England, a mob broke the toll-gates, because they thought an unjust tax was being put upon them. They did not perceive that this small tax for the use of a road would confer upon them innumerable comforts, and double and treble the means of employment.

If there were no road, and no bridge, a man would take six months in finding his way from London to Edinburgh, if indeed he found it at all. He would have to keep the line of the hills, in order that he might come upon the rivers at particular spots, where he would be able to jump over them with ease, or wade through them without danger.

When a man has gone up the bank of a river for twelve miles in one direction, in order to be able to cross it, he may find that, before he proceeds one mile in the line of his journey, he has to go along the bank of another river for twelve miles in the opposite direction; and the courses of the rivers may be so crooked that he is really farther from his journey's end at night than he was in the morning.

He may come to the side of a lake, and not know the end at which the river, too broad and deep for him to cross, runs out; and he may go twenty miles the wrong way, and thus lose forty.

Difficulties such as these are felt by every traveller in an uncivilized country. In reading books of travels, in Africa for instance, we sometimes wonder how it is that the adventurer proceeds a very few miles each day. We forget that he has no roads.

Two hundred years ago—even one hundred years ago—in some places fifty years ago—the roads of England were wholly unfit for general traffic and the conveyance of heavy goods. Pack-horses mostly carried on the communication in the manufacturing districts. The roads were as unfit for moving commodities of bulk, such as coal, wool, and corn, as the sandy roads of Poland were thirty years ago, and as many still are. Mr. Jacob, who went upon the continent to see what stores of wheat existed, found that in many parts the original price of wheat was doubled by the price of land conveyance for a very few miles.

In 1663 the first turnpike act, which was so offensive to some of the people, was carried through Parliament. It was for the repair of the "ancient highway and post-road leading from London to York," which was declared to be "very ruinous, and become almost impassable." This was, on many accounts, one of the most important lines of the country. Let us see in what state it was seventeen years after the passing of the act. In the 'Diary of Ralph Thoresby,' under the date of October, 1680, we have this entry:—"To Ware, twenty-miles from London, a most pleasant road in summer, and as bad in winter, because of the depth of the cart-ruts." Take another road a little later. In December, 1703, Charles III., King of Spain, slept at Petworth on his way from Portsmouth to Windsor, and Prince George of Denmark went to meet him there by desire of the Queen. The distance from Windsor to Petworth is about forty miles. In the relation of the journey given by one of the prince's attendants, he states,—"We set out at six in the morning, by torchlight, to go to Petworth, and did not get out of the coaches (save only when we were overturned or stuck fast in the mire) till we arrived at our journey's end. 'Twas a hard service for the Prince to sit fourteen hours in the coach that day without eating anything, and passing through the worst ways I over saw in my life. We were thrown but once indeed in going, but our coach, which was the leading one, and his Highness's body-coach, would have suffered very much, if the nimble poors of Sussex had not frequently poised it, or supported it with their shoulders, from Godalming almost to Petworth, and the nearer we approached the duke's house the more inaccessible it seemed to be. The last nine miles of the way cost us six hours' time to conquer them." From Horsham, the county-town of Sussex, about the beginning of the reign of George III., the roads were never in such a condition as to allow sheep or cattle to be driven on them to the London market; and consequently, there not being sufficient demand at home to give a remunerating price, the beef and mutton were sold at a rate far below the average to the small population in the country, which was thus isolated from the common channels of demand and supply.

Telford.

In the Highlands of Scotland, at the beginning of the present century, the communication from one district to another was attended with such difficulty and danger, that some of the counties were excused from sending jurors to the circuit to assist in the administration of justice. The poor people inhabiting these districts were almost entirely cut off from intercourse with the rest of mankind. The Highlands were of less advantage to the British empire than the most distant colony. Parliament resolved to remedy the evil; and, accordingly, from 1802 to 1817, the sum of two hundred thousand pounds was laid out in making roads and bridges in these mountainous districts. Mark the important consequences to the people of the Highlands, as described by Mr. Telford, the engineer of the roads:—

> "Since these roads were made accessible, wheelwrights and cartwrights have been established, the plough has been introduced, and improved tools and utensils are used. The plough was not previously used in general; in the interior and mountainous parts they frequently used crooked sticks with iron on them, drawn or pushed along. The moral habits of the great mass of the working classes are changed; they see that they may depend on their own exertions for support. This goes on silently, and is scarcely perceived until apparent by the results. I consider these improvements one of the greatest blessings ever conferred upon any country. About two hundred thousand pounds has been granted in fifteen years. It has been the means of advancing the country at least one hundred years."

There are many parts of Ireland which sustained the same miseries and inconveniences from the want of roads as the Highlands of Scotland did at the beginning of the present century. In 1823 Mr. Nimmo, the engineer, stated to parliament, that the fertile plains of Limerick, Cork, and Kerry, were separated from each other by a deserted country, presenting an impassable barrier between them. This region was the retreat of smugglers, robbers, and culprits of every description; for the tract was a wild, neglected, and deserted country, without roads, culture, or civilization. The government ordered roads to be made through this barren district. We will take one example of the immediate effect of this road-making, as described by a witness before Parliament:—"A hatter, at Castle-island, had a small field through which the new road passed; this part next the town was not opened until 1826. In making arrangements with him for his damages, he said that he ought to make me (the engineer) a present of all the land he had, for that the second year I was at the roads he sold more hats to the people of the mountains alone than he did for seven years before to the high and low lands together. Although he never worked a day on the roads, he got comfort and prosperity by them."

The hatter of Castle-island got comfort and prosperity by the roads, because the man who had to sell and the man who had to buy were brought closer to each other by means of the roads. When there were no roads, the hatter kept his goods upon the shelf, and the labourer in the mountains went without a hat. When the labourer and the hatter were brought together by the roads, the hatter soon sold off his stock, and the manufacturer of hats went to work to produce him a new stock; while the labourer, who found the advantage of having a hat, also went to work to earn more money, that he might pay for another when he should require it. It became a fashion to wear hats, and of course a fashion to work hard, and to save time, to be able to pay for them. Thus the road created industry on both sides,—on the side of the producer of hats and that of the consumer.

Instances such as these of the want of communication between one district and another are now very rare indeed in these islands. But if we look to countries intimately connected with our own, we shall find no lack of examples of a state of commercial intercourse attending a want of roads. The gold-fields of Australia have largely stimulated the export of manufactured goods from Great Britain. One of the colonists at Sydney writes thus to the chief organ of intelligence in England:—"The roads throughout the colony, bad as they were, are now worse than ever. The inland mails cannot run by night, and stick fast and upset in all directions by day. Communication with the interior towns is possible only at enormous cost. The price of conveying a ton of goods from Sydney to Bathurst, about 130 miles, is eight times the freight of the same quantity from London to Sydney. In cost of conveyance London and Liverpool are, in fact, only sixteen miles from Sydney by land, though the distance by sea is 16,000. We here see daily the most striking illustration of the truth that

'Seas but join the regions they divide.'

Cargoes are poured into the seaports with the greatest facility, and then the distribution is suddenly checked. Hence the enormous rents of stores, cessation of demand, and the necessity of forced sales, with the natural consequence—heavy losses to the exporters, who perhaps wonder how trade with Australia can be so unprofitable, scarcely suspecting one of the main causes of its uncertainty. English merchants might do worse than help to open up the internal communications of this continent."

The city of Sydney has a wharfage two miles in extent. The communication from the port to the interior is thus described:—"Imagine the Great Western Railroad, instead of terminating in a splendid station, with every means of conveying and removing goods to roads in every direction, ending suddenly in swamp, forest, and sand, through which, by dint of lashing, and swearing, and unloading, and reloading, a team of bullocks and a dray drag their Manchester goods ten miles *per diem*, at 50*l.* or 80*l.* per ton for the journey. The channel of trade is all that civilization, science, and capital can make it, from the threshold of the Manchester factory to the edge of the Sydney wharf. There it breaks suddenly, and beyond all is primitive, rude, and barbarous in the means of conveyance. The bale of goods last unloaded from the railway train is transferred to the bullock dray, to begin its 'crawl' up the country, costing all its freight from England for every twenty miles. It cannot be otherwise. There are no passable roads."

Modern Syrian Cart.

It is impossible to have a more vivid picture than this of the sudden impediment which the commercial enterprise of one country receives from the want of the commonest means of communication in another. The bullock-cart of Syria, and the Australian bullock-cart, would be useful instruments if they had roads to work in. But there must be general civilization before there are extensive roads. Carts and bullocks are of readier creation than roads. It has taken eighteen centuries to make our English roads, and the Romans, the kings of the world, were our great road-makers, whose works still remain:—

"labouring pioneers,

> A multitude with spades and axes arm'd,
> To lay hills plain, fell woods, or valleys fill,
> Or where plain was raise hill, or overlay
> With bridges rivers proud, as with a yoke."—PARADISE REGAINED.

What the Romans were to England, the colonized English must be to Australia. But the discovery of great natural wealth, the vigour of the race, the intercourse with commercial nations of the old and new world, the free institutions which have been transplanted there without any arbitrary meddling or chilling patronage, will effect in a quarter of a century what the parent people, struggling with ignorant rulers and feeble resources, have been ages in accomplishing.

It is encouraging to all nations to see what we have accomplished in this direction.

In 1839 the turnpike-roads of England and Wales amounted to 21,962 miles, and in Scotland to 3666 miles; while in England and Wales the other highways amounted to 104,772 miles. The turnpike-roads were maintained at a cost of a million a year; and the parish highways at a cost of about twelve hundred thousand pounds. There were at that time nearly eight thousand toll-gates in England and Wales. There had been two thousand miles of turnpike-roads, and ten thousand miles of other highways, added to the number existing in 1814. But the improvements of all our roads during that period had been enormous. Science was brought to bear upon the turnpike lines. Common sense changed their form and re-organized their material. The most beautiful engineering was applied to raise valleys and lower hills. Mountains were crossed with ease; rivers were spanned over by massive piers, or by bridges which hung in the air like fairy platforms. The names of M'Adam and Telford became "household words;" and even parish surveyors, stimulated by example, took thought how to mend their ways.

The Canals of England date only for a hundred years back. The first Act of Parliament for the construction of a canal was passed in 1755. The Duke of Bridgewater obtained his first Act of Parliament in 1759, for the construction of those noble works which will connect his memory with those who have been the greatest benefactors of their country. The great manufacturing prosperity of England dates from this period; and it will be for ever associated with the names of Watt, the improver and almost the inventor of the steam-engine,—of Arkwright, the presiding genius of cotton-spinning,—and of Brindley, the great engineer of canals. In the conception of the vast works which Brindley undertook for the Duke of Bridgewater, there was an originality and boldness which may have been carried further in recent engineering, but which a century ago were the creators of works which were looked upon as marvels. To cut tunnels through hills—to carry mounds across valleys—to build aqueducts over navigable rivers—were regarded then as wild and impracticable conceptions. Another engineer, at Brindley's desire, was called in to give an opinion as to a proposed aqueduct over the river Irwell. He looked at the spot where the aqueduct was to be built, and exclaimed, "I have often heard of castles in the air, but never before was shown the place where any of them were to be erected." Brindley's castle in the air still stands firm; and his example, and that of his truly illustrious employer, have covered our land with many such fabrics, which owe their origin not to the government but to the people.

Brindley's Aqueduct over the Irwell.

The navigable canals of England are more than two thousand miles in length. For the slow transport of heavy goods they hold their place against the competition of railroads, and continue to be important instruments of internal commerce. When railways were first projected it is said that an engineer, being asked what would become of the canals if the new mode of transit were adopted, answered that they would be drained and become the beds of railways. Like many other predictions connected with the last great medium of internal communication, the engineer was wholly mistaken in his prophecy.

The great principle of exchange between one part of this empire and another part, which has ceased to be an affair of restrictions and jealousies, has covered the island with good roads, with canals, and finally with railways. The railway and the steam-carriage have carried the principle of diminishing the price of conveyance, and therefore of commodities, by machinery, to an extent which makes all other illustrations almost unnecessary. A road with a waggon moving on it is a mechanical combination; a canal, with its locks, and towing-paths, and boats gliding along almost without effort, is a higher mechanical combination; a railway, with its locomotive engine, and carriage after carriage dragged along at the rate of thirty or forty miles an hour, is the highest of such mechanical combinations. The force applied upon a level turnpike-road, which is required to move 1800 lbs., if applied to drag a canal-boat will move 55,500 lbs., both at the rate of 2-1/2 miles per hour. But we want economy in time as well as economy in the application of motive power. It has been attempted to apply speed to canal travelling. Up to four miles an hour the canal can convey an equal weight more economically than a railroad; but after a certain velocity is exceeded, that is 13-1/2 miles an hour, the horse on the turnpike-road can drag as much as the canal-team. Then comes in the great advantage of the railroad. The same force that is required to draw 1900 lbs. upon a canal, at a rate above 13-1/2 miles an hour, will draw 14,400 lbs. upon a railway, at the rate of 13-1/2 miles an hour. The producers and consumers are thus brought together, not only at the least cost of transit, but at the least expenditure of time. The road, the canal, and the railway have each their distinctive advantages; and it is worthy of note how they work together. From every railway station there must be a road to the adjacent towns and villages, and a better road than was once thought necessary. Horses are required as much as ever, although mails and post-chaises are no longer the glories of the road; and the post finds its way into every hamlet by the united agency of the road and the railway.

Roger North described a Newcastle railway in 1680:—"Another thing that is remarkable is their way-leaves; for when men have pieces of ground between the colliery and the river, they sell leave to lead coals over their ground; and so dear that the owner of a rood of ground will expect 20*l.* per annum for this leave. The manner of the carriage is by laying rails of timber, from the colliery down to the river, exactly straight and parallel; and bulky carts are made with four rowlets fitting these rails; whereby the carriage is so easy that the horse will draw down four or five chaldron of coals, and is an immense benefit to the coal-merchant." Who would have thought that this contrivance would have led to no large results till a hundred and fifty years had passed away? Who could have believed that "the rails of timber, exactly straight and parallel," and the "bulky carts with four rowlets exactly fitting the rails," would have changed the face, and to a great degree the destinies, of the world?

If we add to the road, the canal, and the railway, the steam-boat traffic of our own coasts, we cannot hesitate to believe that the whole territory of Great Britain and Ireland is more compact, more closely united, more accessible, than was a single county two centuries ago. It may be said, without exaggeration, that it would now be impossible for a traveller in England to set himself down in any accessible situation where the post from London would not reach him in twelve hours. When the first edition of the 'Results of Machinery' was published in 1831, we said that the post from London would reach any part of England in three days; and that, "fifty years before, such a quickness of communication would have been considered beyond the compass of human means." In twenty-four years we have so diminished the practical amount of distance between one part of Great Britain and another, that the post from London to Aberdeen is carried five hundred and forty miles in little more than twenty hours. It is this wonderful rapidity of communication, in connection with the cheapness of postage, which has multiplied letters five-fold since 1839, when the penny rate was introduced. In that year the number of chargeable and franked letters distributed in the United Kingdom was eighty-two millions; in 1853 it was four hundred and ten millions.

Locomotive-Engine Factory.

The annual returns of our railways furnish some of the most astounding figures of modern statistics. On the 1st of January 1854 there were open in England 5811 miles of railway; in Scotland, 995 miles; in Ireland, 834 miles. In 1853 there were one hundred and two million passengers conveyed, who travelled one billion five hundred million miles, being an average of nearly fifteen miles to each passenger. In England considerably less than one-half of the passengers were by penny-a-mile and other third-class trains; in Ireland one-half; and in Scotland two-thirds. The receipts from goods traffic exceed those of the passenger traffic in England and Scotland, but are less in Ireland. These are indeed wonderful results from a system which was wholly experimental twenty-five years ago.

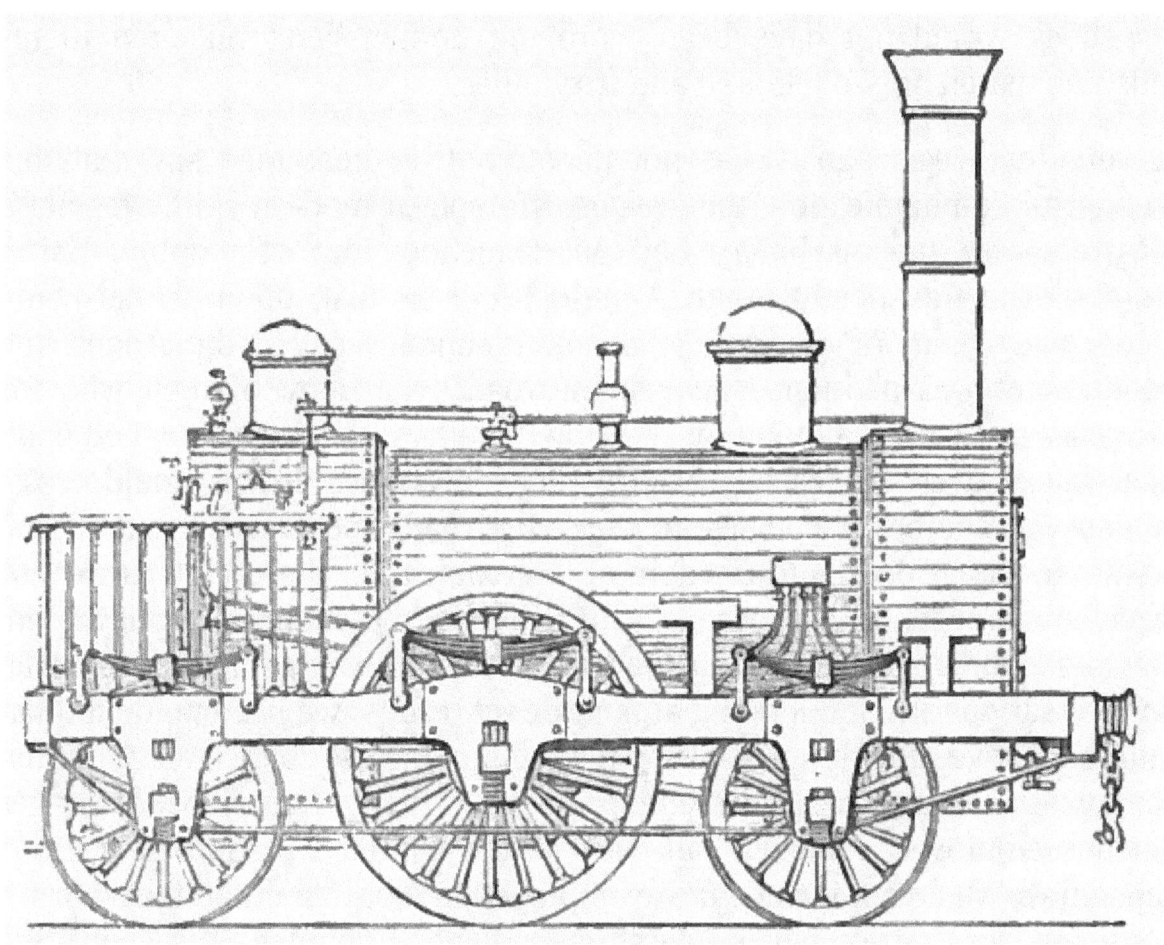

Railway Locomotive.

When William Hutton, in the middle of last century, started from Nottingham (where he earned a scanty living as a bookbinder) and walked to London and back for the purpose of buying tools, he was nine days from home, six of which were spent in going and returning. He travelled on foot, dreading robbers, and still more dreading the cost of food and lodging at public-houses. His whole expenses during this toilsome expedition were only ten shillings and eight pence; but he contented himself with the barest necessaries, keeping the money for his tools sewed up in his shirt-collar. If William Hutton had lived in these days, he would, upon sheer principles of economy, have gone to London by the Nottingham train at a cost of twenty shillings for his transit, in one forenoon, and returned in another. The twenty shillings would have been sacrificed for his conveyance, but he would have had a week's labour free to go to work with his new tools; he need not have sewed his money in his shirt-collar for fear of thieves; and his shoes would not have been worn out and his feet blistered in his toilsome march of two hundred and fifty miles.

A very few years ago it was not uncommon to hear men say that this wonderful communication, the greatest triumph of modern skill, was not a blessing;—for the machinery had put somebody out of employ. Baron Humboldt, a traveller in South America, tells us that, upon a road being made over a part of the great chain of mountains called the Andes, the government was petitioned against the road by a body of men who for centuries had gained a living by carrying travellers in baskets strapped upon their backs over the fearful rocks, which only these guides could cross. Which was the better course—to make the road, and create the thousand employments belonging to freedom of intercourse, for these very carriers of travellers, and for all other men; or to leave the mountains without a road, that the poor guides might gain a premium for risking their lives in an unnecessary peril? But, looking at their direct results, we have no doubt that railroads have greatly multiplied the employments connected with the conveyance of goods and passengers. In 1853 there were eighty thousand persons employed upon the railroads of the United Kingdom in various capacities. We do not include those employed in working upon lines that are not open for traffic, which class in England amounted to twenty-five thousand persons in 1853. But the indirect occupations called into activity by railroads are so numerous as to defy all attempts at calculating the numbers engaged in them. No doubt many occupations were changed by railroads;—there were fewer coachmen, guards, postboys, waggoners, and others, on such a post-road as that from London to York. But it is equally certain that throughout the kingdom there are far more persons employed in conducting the internal communication of the country, effecting that great addition to its productive powers, without which all other production would languish and decay. The census returns of 1851 give the number of three hundred and eighty-six thousand males so employed, including those engaged on our rivers, canals, and coast traffic.

Reindeer.

The vast extension, and the new channels, of our foreign commerce have been greatly affected by the prodigious facilities of our internal communication. They have created, in a measure, special departments of industry, which can be most advantageously pursued in particular localities; but which railways and steam-vessels have united with the whole kingdom, with its colonies, with the habitable globe. The reindeer connects the Laplander with the markets of Sweden, and draws his sledge over the frozen wilds at a speed and power of continuance only rivalled by the locomotive. The same beneficent Providence which has given this animal to the inhabitant of the polar regions,—not only for food, for clothing, but for transport to associate him with some civilization,—has bestowed upon us the mighty power of steam, to connect us with the entire world, from which we were once held to be wholly separated.

CHAPTER XIV.

> Houses—The pyramids—Mechanical power—Carpenters' tools—American machinery for building—Bricks—Slate—Household fittings and furniture—Paper-hangings—Carpets—Glass—Pottery—Improvements effected through the reduction or repeal of duties on domestic requirements.

The beaver builds his huts with the tools which nature has given him. He gnaws pieces of wood in two with his sharp teeth, so sharp that the teeth of a similar animal, the agouti, form the only cutting-tool which some rude nations possess. When the beavers desire to move a large piece of wood, they join in a body to drag it along.

Man has not teeth that will cut wood: but he has reason, which directs him to the choice of much more perfect tools.

Pyramid and sphinx

Some of the great monuments of antiquity, such as the pyramids of Egypt, are constructed of enormous blocks of stone brought from distant quarries. We have no means of estimating, with any accuracy, the mechanical knowledge possessed by the people engaged in these works. It was, probably, very small, and, consequently, the human labour employed in such edifices was not only enormous in quantity, but exceedingly painful to the workmen. The Egyptians, according to Herodotus, a Greek writer who lived two thousand five hundred years ago, hated the memory of the kings who built the pyramids. He tells us that the great pyramid occupied a hundred thousand men for twenty years in its erection, without counting the workmen who were employed in hewing the stones, and in conveying them to the spot where the pyramid was built. Herodotus speaks of this work as a torment to the people; and doubtless the labour engaged in raising huge masses of stone, that was extensive enough to employ a hundred thousand men for twenty years, which is equal to two millions of men for one year, must have been fearfully tormenting without machinery, or with very imperfect machinery. It has been calculated that about half the steam-engines of England, worked by thirty-six thousand men, would raise the same quantity of stones from the quarry, and elevate them to the same height as the great pyramid, in the short time of eighteen hours. The people of Egypt groaned for twenty years under this enormous work. The labourers groaned because they were sorely taxed; and the rest of the people groaned because they had to pay the labourers. The labourers lived, it is true, upon the wages of their labour, that is, they were paid in food—kept like horses—as the reward of their work. Herodotus says that it was recorded on the pyramid that the onions, radishes, and garlic which the labourers consumed, cost sixteen hundred talents of silver: an immense sum, equivalent to several million pounds. But the onions, radishes, and garlic, the bread, and clothes of the labourer, were wrung out of the profitable labour of the rest of the people. The building of the pyramid was an unprofitable labour. There was no immediate or future source of produce in the pyramid; it produced neither food, nor fuel, nor clothes, nor any other necessary. The labour of a hundred thousand men for twenty years, stupidly employed upon this monument, without an object beyond that of gratifying the pride of the tyrant who raised it, was a direct tax upon the profitable labour of the rest of the people.

> "Instead of useful works, like nature great,
> Enormous cruel wonders crush'd the land."

But admitting that it is sometimes desirable for nations and governments to erect monuments which are not of direct utility,—which may have an indirect utility in recording the memory of great exploits, or in producing feelings of reverence or devotion,—it is clearly an advantage that these works, as well as all other works, should be performed in the cheapest manner; that is, that human labour should derive every possible assistance from mechanical aid. We will give an illustration of the differences of the application of mechanical aid in one of the first operations of building, the moving a block of stone. The following statements are the result of actual experiment upon a stone weighing ten hundred and eighty pounds.

Portland Quarry.

To drag this stone along the smoothed floor of the quarry required a force equal to seven hundred and fifty-eight pounds. The same stone dragged over a floor of planks required six hundred and fifty-two pounds. The same stone placed on a platform of wood, and dragged over the same floor of planks, required six hundred and six pounds. When the two surfaces of wood were soaped as they slid over each other, the force required to drag the stone was reduced to one hundred and eighty-two pounds. When the same stone was placed upon rollers three inches in diameter, it required, to put it in motion along the floor of the quarry, a force only of thirty-four pounds; and by the same rollers upon a wooden floor, a force only of twenty-eight pounds. Without any mechanical aid, it would require the force of four or five men to set that stone in motion. With the mechanical aid of two surfaces of wood soaped, the same weight might be moved by one man. With the more perfect mechanical aid of rollers, the same weight might be moved by a very little child.

From these statements it must be evident that the cost of a block of stone very much depends upon the quantity of labour necessary to move it from the quarry to the place where it is wanted to be used. We have seen that with the simplest mechanical aid labour may be reduced sixty-fold. With more perfect mechanical aid, such as that of water-carriage, the labour may be reduced infinitely lower. Thus, the streets of London are paved with granite from Scotland at a moderate expense.

The cost of timber, which enters so largely into the cost of a house, is in a great degree the cost of transport. In countries where there are great forests, timber-trees are worth nothing where they grow, except there are ready means of transport. In many parts of North America, the great difficulty which the people find is in clearing the land of the timber. The finest trees are not only worthless, but are a positive incumbrance, except when they are growing upon the banks of a great river; in which case the logs are thrown into the water, or formed into rafts, being floated several hundred miles at scarcely any expense. The same stream which carries them to a seaport turns a mill to saw the logs into planks; and when sawn into planks the timber is put on shipboard, and carried to distant countries where timber is wanted. Thus mechanical aid alone gives a value to the timber, and by so doing employs human labour. The stream that floats the tree, the sawing-mill that cuts it, the ship that carries it across the sea, enable men profitably to employ themselves in working it. Without the stream, the mill, and the ship, those men would have no labour, because none could afford to bring the timber to their own doors.

Timber Rafts of the Tyrol.

What an infinite variety of machines, in combination with the human hand, is found in a carpenter's chest of tools! The skilful hand of the workman is the *power* which sets these machines in motion; just as the wind or the water is the power of a mill, or the elastic force of vapour the power of a steam-engine. When Mr. Boulton, the partner of the great James Watt, waited upon George III. to explain one of the improvements of the steam-engine which they had effected, the king said to him, "What do you sell, Mr. Boulton?" and the honest engineer answered, "What kings, sire, are all fond of—*power*." There are people at Birmingham who let out *power*, that is, there are people who have steam-engines who will lend the use of them, by the day or the hour, to persons who require that saving of labour in their various trades; so that a person who wants the strength of a horse, or half a horse, to turn a wheel for grinding, or for setting a lathe in motion, hires a room, or part of a room, in a mill, and has just as much as he requires. The *power* of a carpenter is in his hand, and the machines moved by that power are in his chest of tools. Every tool which he possesses has for its object to reduce labour, to save material, and to ensure accuracy—the objects of all machines. What a quantity of waste both of time and stuff is saved by his foot-rule! and when he chalks a bit of string and stretches it from one end of a plank to the other, to jerk off the chalk from the string, and thus produce an unerring line upon the face of the plank, he makes a little machine which saves him great labour. Every one of his hundreds of tools, capable of application to a vast variety of purposes, is an invention to save labour. Without some tool the carpenter's work could not be done at all by the human hand. A knife would do very laboriously what is done very quickly by a hatchet. The labour of using a hatchet, and the material which it wastes, are saved twenty times over by the saw. But when the more delicate operations of carpentry are required—when the workman uses his planes, his rabbet-planes, his fillisters, his bevels, and his centre-bits—what an infinitely greater quantity of labour is economized, and how beautifully that work is performed, which, without them, would be rough and imperfect! Every boy of mechanical ingenuity has tried with his knife to make a boat; and with a knife only it is the work of weeks. Give him a chisel, and a gouge, and a vice to hold his wood, and the little boat is the work of a day. Let a boy try to make a round wooden box, with a lid, having only his knife, and he must be expert indeed to produce anything that will be neat and serviceable. Give him a lathe and chisels, and he will learn to make a tidy box in half an hour. Nothing but absolute necessity can render it expedient to use an imperfect tool instead of a perfect. We sometimes see exhibitions of carving, "all done with the common penknife." Professor Willis has truly said, with reference to such weak boasting, "So far from admiring, we should pity the vanity and folly of such a display; and the more, if the work should show a natural aptitude in the workman: for it is certain that, if he has made good work with a bad tool, he would make better with a good one."

Boulton.

The Emperor Maximilian, at the beginning of the sixteenth century, ordered a woodcut to be engraved that should represent the carpentry operations of his time and country. This prince was, no doubt, proud of the advance of Germany in the useful arts. If the President of the United States were thus to record the advance of the republic of which he is the chief, he would show us his saw-mills and his planing-mills. The German carpenters, as we see, are reducing a great slab of wood into shape by the saw and the adze. The Americans have planing-mills, with cutters that make 4000 revolutions, and which plane boards eighteen feet long at the rate of fifty feet, per minute;

and while the face of the board is planed, it is tongued and grooved at the same time—that is, one board is made to fit closely into another. But the Americans carry machinery much farther into the business of carpentry. Mr. Whitworth tells us that "many works in various towns are occupied exclusively in making doors, window-frames, or staircases, by means of self-acting machinery, such as planing, tenoning, morticing, and jointing machines. They are able to supply builders with various parts of the wood-work required in building at a much cheaper rate than they can produce them in their own workshops without the aid of such machinery."

Carpenters and their tools. (From an old German woodcut.)

By the use of those machines we are told that twenty men can make panelled doors at the rate of a hundred a day—that is, one man can make five doors. A panelled door is a very expensive part of an English house; and so are window-frames and staircases. If doors and windows and staircases can be made cheaper, more houses and better houses will be built; and thus more carpenters will be employed in building than if those parts of a house were made by hand. The same principle applies to machines as to tools. If carpenters had not tools to make houses, there would be few houses made; and those that were made would be as rough as the hut of the savage who has no tools. The people would go without houses, and the carpenter would go without work,—to say nothing of the people, who would also go without work, that now make tools for the carpenter.

We build in this country more of brick than of stone, because brick-earth is found almost everywhere, and stone fit for building is found only in particular districts. Bricks used to pay the state a duty of five shillings and ten pence a thousand; and yet at the kilns they were to be bought under forty shillings a thousand, which is less than a halfpenny apiece. The government wisely resolved, in 1850, to repeal the excise-duty on bricks. In 1845 the duty on glass was repealed. In 1847 the timber-duties were reduced; and in 1848 they were further reduced. The ever-present necessities of the people—the absolute want of house-accommodation for a population increasing so rapidly—rendered it a paramount duty of the government no longer to let tax interfere with the cheap building of houses. Every invention that adds to cheapness acts in the same direction; for although the direct taxes cease to press upon the various trades of building, the constant demand keeps bricks and timber at a price almost as high as before the removal or mitigation of the tax. But bricks, regarded as the production of a vast amount of labour, are intrinsically cheap. And why? Because they are made by what is truly machinery; as they were made three thousand years ago by the Egyptians.

The clay is ground in a horse-mill; the wooden mould, in which every brick is made singly, is a copying machine. One brick is exactly like another brick. Every brick is of the form of the mould in which it is made. Without the mould the workman could not make the brick of uniform dimensions; and without this uniformity the after labour of putting bricks together would be greatly increased. Without the mould the workman could not form the bricks quickly;—his own labour would be increased ten-fold. The simple machine of the mould not only gives employment to a great many brickmakers who would not be employed at all, but also to a great many bricklayers who would also want employment if the original cost of production were so enormously increased.

Egyptian labour in the brick-field.

There is another material for building which was little used at the beginning of the century. The consumption of slate in London alone was, in 1851, from thirty thousand to forty thousand tons per annum. The quarries of Wales principally supply this immense quantity; but some slates are shipped from Lancashire and Westmorland, and from Scotland and Ireland. In the production of this one material, eight thousand quarriers are employed in Great Britain. Slates are not only used for roofing houses, but in slabs for cisterns and chimney-pieces. The great increase of the supply of water to houses by machinery led to a demand for a safer and cheaper material than lead for cisterns; and slate supplied the want.

How great a variety of things are contained in an ironmonger's shop! Half his store consists of tools of one sort or another to save labour; and the other half consists of articles of convenience or elegance most perfectly adapted to every possible want of the builder or the maker of furniture. The uncivilized man is delighted when he obtains a nail,—any nail. A carpenter and joiner, who supply the wants of a highly civilized community, are not satisfied unless they have a choice of nails, from the finest brad to the largest clasp-nail. A savage thinks a nail will hold two pieces of wood together more completely than anything else in the world. It is seldom, however, that he can afford to put it to such a use. If it is large enough, he makes it into a chisel. An English joiner knows that screws will do the work more perfectly in some cases than any nail; and therefore we have as great a variety of screws as of nails. The commonest house built in England has hinges, and locks, and bolts. A great number are finished with ornamented knobs to door-handles, with bells and bell-pulls, and a thousand other things that have grown up into necessities, because they save domestic labour, and add to domestic comfort. And many of these things really are necessities. M. Say, a French writer, gives us an example of this; and as his story is an amusing one, besides having a moral, we may as well copy it:—

"Being in the country," says he, "I had an example of one of those small losses which a family is exposed to through negligence. For the want of a latchet of small value, the wicket of a barn-yard leading to the fields was often left open. Every one who went through drew the door to: but as there was nothing to fasten the door with, it was always left flapping; sometimes open, and sometimes shut. So the cocks and hens, and the chickens, got out, and were lost. One day a fine pig got out, and ran off into the woods; and after the pig ran all the people about the place,—the gardener, and the cook, and the dairymaid. The gardener first caught sight of the runaway, and, hastening after it, sprained his ankle; in consequence of which the poor man was not able to get out of the house again for a fortnight. The cook found, when she came back from pursuing the pig, that the linen she had left by the fire had fallen down and was burning; and the dairymaid having, in her haste, neglected to tie up the legs of one of her cows, the cow had kicked a colt, which was in the same stable, and broken its leg. The gardener's lost time was worth twenty crowns, to say nothing of the pain he suffered. The linen which was burned, and the colt which was spoiled, were worth as much more. Here, then, was caused a loss of forty crowns, as well as much trouble, plague, and vexation, for the want of a latch which would not have cost three-pence." M. Say's story is one of the many examples of the truth of the old proverb—"for want of a nail the shoe was lost, for want of a shoe the horse was lost, for want of a horse the man was lost."

Nearly all the great variety of articles in an ironmonger's shop are made by machinery. Without machinery they could not be made at all, or they would be sold at a price which would prevent them being commonly used. Some of the finer articles, such as a Bramah lock, or a Chubb's lock, could not be made at all, unless machinery had been called in to produce that wonderful accuracy, through which no one of a hundred thousand locks and keys shall be exactly like another lock and key. With machinery, the manufacture of ironmongery employs large numbers of artisans who would be otherwise unemployed. There are hundreds of ingenious men at Birmingham who go into business with a capital acquired by their savings as workmen, for the purpose of manufacturing some one single article used in finishing a house, such as the knob of a lock. All the heavy work of their trade is done by machinery. The cheapness of the article creates workmen; and the savings of the workmen accumulate capital to be expended in larger works, and to employ more workmen.

The furniture of a house, some may say—the chairs, and tables, and bedsteads—is made nearly altogether by hand. True. But tools are machines; and further, we owe it to what men generally call machinery, that such furniture, even in the house of a very poor man, is more tasteful in its construction, and of finer material, than that possessed by a nobleman a hundred years ago. How is this? Machinery (that is ships) has brought us much finer woods than we grow ourselves; and other machinery (the sawing-mill) has taught us how to render that fine wood very cheap, by economising the use of it. At a veneering-mill, that is, a mill which cuts a mahogany log into thin plates, much more delicately and truly, and in infinitely less time, than they could be cut by the hand, two hundred and forty square feet of mahogany are cut by one circular saw in one hour. A veneer, or thin plate, is cut off a piece of mahogany, six feet six inches long, by twelve inches wide, in twenty-five seconds. What is the consequence of this? A mahogany table is made almost as cheap as a deal one; and thus the humblest family in England may have some article of mahogany, if it be only a tea-caddy. And let it not be said that deal furniture would afford as much happiness; for a desire for comfort, and even for some degree of elegance, gives a refinement to the character, and, in a certain degree, raises our self-respect. Diogenes, who is said to have lived in a tub, was a great philosopher; but it is not necessary to live in a tub to be wise and virtuous. Nor is that the likeliest plan for becoming so. The probability is, that a man will be more wise and virtuous in proportion as he strives to surround himself with the comforts and decent ornaments of his station.

It is a circumstance worthy to be borne in mind by all who seek the improvement of the people, that whatever raises not only the standard of comfort, but of taste, has direct effects of utility which might not at first be perceived. We will take the case of paper-hangings. Their very name shows that they were a substitute for the arras, or hangings, of former times, which were suspended from the ceilings to cover the imperfections of the walls. This was the case in the houses of the rich. The poor man in his hut had no such device, but must needs "patch a hole to keep the wind away." Till 1830, what, in the language of the excise, was called stained paper, was enormously dear, for a heavy tax greatly impeded its production. When it was dear, many walls were stencilled or daubed over with a rude pattern. The paper-hangings themselves were not only dear, but offensive to the eye, from their want of harmony in colour and of beauty in design. The old papers remained on walls for half a century; and it was not till paper-hangings became a penny a yard, or even a halfpenny, that the landlord or tenant of a small house thought of re-papering. The eye at length got offended by the dirty and ugly old paper. The walls were recovered with neat patterns. But what had offended the eye had been prejudicial to the health. The old papers, that were saturated with damp from without and bad air from within, were recipients and holders of fever. When the bed-room became neat it also became healthful. The duty on paper was 1-3/4*d.* per yard, when the paper-hanger used to paste together yard after yard, made by hand at the paper-mill, and stamped by block. The paper-machine which gave long rolls of paper enabled hangings to be printed by cylinder, as calico is printed. The absence of tax, and the improvement of the manufacture by machinery, have enabled every man to repaper his filthy and noxious room for almost as little as its whitewashing or colouring will cost him.

Look, again, at the carpet. Contrast it in all its varieties, from the gorgeous Persian to the neat Kidderminster, with the rushes of our forefathers, amidst which the dogs hunted for the bones that had been thrown upon the floor. The clean rushes were a rare luxury, never thought of but upon some festive occasion. The carpet manufacture was little known in England at the beginning of the last century; as we may judge from our still calling one of the most commonly-woven English carpets by the name of "Brussels." There are twelve thousand persons now employed in the manufacture of carpets in Great Britain. The Scotch carpet is the cheapest of the produce of the carpet-loom; and it may be sufficient to show the connection of machinery with the commonest as well as the finest of these productions by an engraving of the loom. One of the most beautiful inventions of man, the Jacquard apparatus (so called from the name of its inventor), is extensively used in every branch of the carpet manufacture.

Scotch carpet-loom.

Let us see what mechanical ingenuity can effect in producing the most useful and ornamental articles of domestic life from the common earth which may be had for digging. Without chemical and mechanical skill we should neither have Glass nor Pottery; and without these articles, how much lowered beneath his present station, in point of comfort and convenience, would be the humblest peasant in the land!

The cost of glass is almost wholly the wages of labour, as the materials are very abundant, and may be said to cost almost nothing; and glass is much more easily worked than any other substance.

Hard and brittle as it is, it has only to be heated, and any form that the workman pleases may be given to it. It melts; but when so hot as to be more susceptible of form than wax or clay, or anything else that we are acquainted with, it still, retains a degree of toughness and capability of extension superior to that of many solids, and of every liquid; when it has become red-hot all its brittleness is gone, and a man may do with it as he pleases. He may press it into a mould; he may take a lump of it upon the end of an iron tube, and, by blowing into the tube with his mouth (keeping the glass hot all the time), he may swell it out into a hollow ball. He may mould that ball into a bottle; he may draw it out lengthways into a pipe; he may cut it open into a cup; he may open it with shears, whirl it round with the edge in the fire, and thus make it into a circular plate. He may also roll it out into sheets, and spin it into threads as fine as a cobweb. In short, so that he keeps it hot, and away from substances by which it may be destroyed, he can do with it just as he pleases. All this, too, may be done, and is done with large quantities every day, in less time than any one would take to give an account of it. In the time that the readiest speaker and clearest describer were telling how one quart bottle is made, an ordinary set of workmen would make some dozens of bottles.

But though the materials of glass are among the cheapest of all materials, and the substance the most obedient to the hand of the workman, there is a great deal of knowledge necessary before glass can be made. It can be made profitably only at large manufactories, and those manufactories must be kept constantly at work night and day.

Glass does not exist in a natural form in many places. The sight of native crystal, probably, led men to think originally of producing a similar substance by art. The fabrication of glass is of high antiquity. The historians of China, Japan, and Tartary speak of glass manufactories existing there more than two thousand years ago. An Egyptian mummy two or three thousand years old, which was exhibited in London, was ornamented with little fragments of coloured glass. The writings of Seneca, a Roman author who lived about the time of our Saviour, and of St. Jerome, who lived five hundred years afterwards, speak of glass being used in windows. It is recorded that the Prior of the convent of Weymouth, in Dorsetshire, in the year 674, sent for French workmen to glaze the windows of his chapel. In the twelfth century the art of making glass was known in this country. Yet it is very doubtful whether glass was employed in windows, excepting those of churches and the houses of the very rich, for several centuries afterwards; and it is quite certain that the period is comparatively recent, as we have shown,[21] when glass windows were used for excluding cold and admitting light in the houses of the great body of the people, or that glass vessels were to be found amongst their ordinary conveniences. The manufacture of glass in England now employs twelve thousand people, because the article, being cheap, is of universal use. The government has wisely taken off the duty on glass; and as the article becomes still cheaper, so will the people employed in its manufacture become more numerous.

Machinery, as we commonly understand the term, is not much employed in the manufacture of glass; but chemistry, which saves as much labour as machinery, and performs work which no machinery could accomplish, is very largely employed. The materials of which glass is made are sand, or earth, and vegetable matter, such as kelp or burnt seaweed, which yield alkali. For the finest glass, sand is brought from great distances, even from Australia. These materials are put in a state of fusion by the heat of an immense furnace. It requires a red heat of sixty hours to prepare the material of a common bottle. Nearly all glass, except glass for mirrors, is what is called blown. The machinery is very simple, consisting only of an iron pipe and the lungs of the workman; and the process is perfected in all its stages by great subdivision of labour, producing extreme neatness and quickness in all persons employed in it. For instance, a wine-glass is made thus:—One man (the blower) takes up the proper quantity of glass on his pipe, and blows it to the size wanted for the bowl; then he whirls it round on a reel, and draws out the stalk. Another man (the footer) blows a smaller and thicker ball, sticks it to the end of the stalk of the blower's glass, and breaks his pipe from it. The blower opens that ball, and whirls the whole round till the foot is formed. Then a boy dips a small rod in the glass-pot, and sticks it to the very centre of the foot. The blower, still turning the glass round, takes a bit of iron, wets it in his mouth, and touches the ball at the place where he wishes the mouth of the glass to be. The glass separates, and the boy takes it to the finisher, who turns the mouth of it; and, by a peculiar swing that he gives it round his head, makes it perfectly circular, at the same time that it is so hardened as to be easily snapped from the rod. Lastly, the boy takes it on a forked iron to the annealing furnace, where it is cooled gradually.

Glass-cutting.

All these operations require the greatest nicety in the workmen; and would take a long time in the performance, and not be very neatly done after all, if they were all done by one man. But the quickness with which they are done by the division of labour is perfectly wonderful.

The cheapness of glass for common use, which cheapness is produced by chemical knowledge and the division of labour, has set the ingenuity of man to work to give greater beauty to glass as an article of luxury. The employment of sharp-grinding wheels, put in motion by a treadle, and used in conjunction with a very nice hand, produces *cut* glass. Cut glass is now comparatively so cheap, that scarcely a family of the middle ranks is without some beautiful article of this manufacture.

Sheet-glass making.

But the repeal of the duty on glass, and of the tax upon windows, has had the effect of improving the architecture of our houses to a degree which no one would have thought possible who had not studied how the operation of a tax impedes production. We have now plate-glass of the largest dimensions, giving light and beauty to our shops; and sheet-glass, nearly as effective as plate, adorning our private dwellings. Sheet-glass, in the making of which an amount of ingenuity is exercised which would have been thought impossible in the early stages of glass-making, is doing for the ordinary purposes of building what plate-glass did formerly for the rich. A portion of melted glass, weighing twelve or fourteen pounds, is, by the exercise of this skill, converted into a ball, and then into a cylinder, and then into a flat plate; and thus two crystal palaces have been built, which have consumed as much glass, weight by weight, as was required for all the houses in one-fourth of the area of Great Britain in the beginning of the century.

Plate-glass Factory.

There are two kinds of pottery—common potters' ware, and porcelain. The first is a pure kind of brick; and the second a mixture of very fine brick and glass. Almost all nations have some knowledge of pottery; and those of the very hot countries are sometimes satisfied with dishes formed by their fingers without any tool, and dried by the heat of the sun. In England pottery of every sort, and in all countries good pottery, must be baked or burnt in a kiln of some kind or other.

Vessels for holding meat and drink are almost as indispensable as the meat and drink themselves; and the two qualities in them that are most valuable are, that they shall be cheap, and easily cleaned. Pottery, as it is now produced in England, possesses both of these qualities in the very highest degree. A white basin, having all the useful properties of the most costly vessels, may be purchased for twopence at the door of any cottage in England. There are very few substances used in human food that have any effect upon these vessels; and it is only rinsing them in hot water, and wiping them with a cloth, and they are clean.

The making of an earthen bowl would be to a man who made a first attempt no easy matter. Let us see how it is done so that it can be carried two or three hundred miles and sold for twopence, leaving a profit to the maker, and the wholesale and retail dealer.

The common pottery is made of pure clay and pure flint. The flint is found only in the chalk counties, and the fine clays in Devonshire and Dorsetshire; so that, with the exception of some clay for coarse ware, the materials out of which the pottery is made have to be carried from the South of England to Staffordshire, where the potteries are situated.

The great advantage that Staffordshire possesses is abundance of coal to burn the ware and supply the engines that grind the materials.

The clay is worked in water by various machinery till it contains no single piece large enough to be visible to the eye. It is like cream in consistence. The flints are burned. They are first ground in a mill, and then worked in water in the same manner as the clay, the large pieces being returned a second time to the mill.

When both are fine enough, one part of flint is mixed with five or six of clay; the whole is worked to a paste, after which it is kneaded either by the hands or a machine; and when the kneading is completed, it is ready for the potter.

He has a little wheel which lies horizontally. He lays a portion of clay on the centre of the wheel, puts one hand, or finger if the vessel is to be a small one, in the middle, and his other hand on the outside, and, as the wheel turns rapidly round, draws up a hollow vessel in an instant. With his hands, or with very simple tools, he brings it to the shape he wishes, cuts it from the wheel with a wire, and a boy carries it off. The potter makes vessel after vessel, as fast as they can be carried away.

The English Potter.

The potter's wheel is an instrument of the highest antiquity. In the book of Ecclesiasticus we read—"So doth the potter, sitting at his work, and turning the wheel about with his feet, who is always carefully set at his work, and maketh all his work by number: he fashioneth the clay with his arm, and boweth down his strength before his feet; he applieth himself to lead it over, and is diligent to make clean the furnace."—(c. xxxix., v. 29, 30.) At the present day the oriental potter stands in a pit, in which the lower machinery of his wheel is placed. He works as the potter of the ancient Hebrews.

As the potter produces the vessels they are partially dried; after which they are turned on a lathe and smoothed with a wet sponge when necessary. Only round vessels can be made on the wheel; those of other shapes are made in moulds of plaster. Handles and other solid parts are pressed in moulds, and stuck on while they and the vessels are still wet.

Potter's wheel of modern Egypt.

The vessels thus formed are first dried in a stove, and, when dry, burnt in a kiln. They are in this state called biscuit. If they are finished white, they are glazed by another process. If they are figured, the patterns are engraved on copper, and printed on coarse paper rubbed with soft soap. The ink is made of some colour that will stand the fire, ground with earthy matter. These patterns are moistened and applied to the porous biscuit, which absorbs the colour, and the paper is washed off, leaving the pattern on the biscuit.

Moulds for porcelain, and casts.

The employment of machinery to do all the heavy part of the work, the division of labour, by which each workman acquires wonderful dexterity in his department, and the conducting of the whole upon a large scale, give bread to a vast number of people, make the pottery cheap, and enable it to be sold at a profit in almost every market in the world. It is not ninety years since the first pottery of a good quality was extensively made in England; and before that time what was used was imported,—the common ware from Delft, in Holland (from which it acquired its name), and the porcelain from China.

Mill-room, where the Ingredients for Pottery are mixed.

The history of the manufacture of porcelain affords us two examples of persevering ingenuity—of intense devotion to one object—which have few parallels in what some may consider the higher walks of art. Palissy and Wedgwood are names that ought to be venerated by every artisan. The one bestowed upon France her manufacture of porcelain, so long the almost exclusive admiration of the wealthy and the tasteful. The other gave to England her more extensive production of earthenware, combining with great cheapness the imitation of the most beautiful forms of ancient art. The potteries of Staffordshire may be almost said to have been created by Josiah Wedgwood. In his workshops we may trace the commencement of a system of improved design which made his ware so superior to any other that had been produced in Europe for common uses. In other branches of manufacture this system found few imitators; and we were too long contented, in our textile fabrics especially, with patterns that were unequalled for ugliness—miserable imitations of foreign goods, or combinations of form and colour outraging every principle of art. We have seen higher things attempted in the present day; but for the greater part of a century the wares of the Staffordshire potter were the only attempts to show that taste was as valuable a quality in association with the various articles which are required for domestic use, as good materials and clean workmanship. It was long before we discovered that taste had an appreciable commercial value.

Wedgwood.

CHAPTER XV.

Dwellings of the people—Oberlin—The Highlander's candlesticks—Supply of water—London waterworks—Street-lights—Sewers.

It is satisfactory to observe that the increase of houses has kept pace with the increase of population. In 1801, in Great Britain, there was a population of ten million five hundred thousand persons, and one million eight hundred thousand inhabited houses. In 1851 there were twenty million eight hundred thousand persons, and three million eight hundred thousand inhabited houses. The numbers, in each case, had, as nearly as may be, doubled.

But it is not equally satisfactory to know that the improvement in the quality of the houses in which the great body of those who labour for wages abide is not commensurate with the increase in their quantity. It is not fitting, that, whilst the general progress of science is raising, as unquestionably it is raising, the average condition of the people—and that whilst education is going forward, slowly indeed, but still advancing—the bulk of those so progressing should be below their proper standard of physical comfort, from the too common want of decent houses to surround them with the sanctities of home.

In the great business of the improvement of their dwellings the working-men require leaders—not demagogues, whose business is to subvert, and not to build up—but leaders like the noble pastor, Oberlin, who converted a barren district into a fruitful, by the example of his unremitting energy. This district was cut off from the rest of the world by the want of roads. Close at hand was Strasburg, full of all the conveniences of social life. There was no money to make roads—but there was abundant power of labour. There were rocks to be blasted, embankments to be raised, bridges to be built. The undaunted clergyman took a pickaxe, and went to work himself. He worked alone, till the people were ashamed of seeing him so work. They came at last to perceive that the thing was to be done, and that it was worth the doing. In three years the road was made. If there were an Oberlin to lead the inhabitants of every filthy street, and the families of every wretched house, to their own proper work of improvement, a terrible evil would be soon removed, which is as great an impediment to the productive powers of a country, and therefore to the happiness of its people, as the want of ready communication, or any other appliance of civilization. The enormity of the evil would be appalling, if the capability of its removal in some degree were not equally certain.

Whatever a government may attempt—whatever municipalities or benevolent associations—there can be nothing so effectual in the upholding to a proper mark the domestic comfort of the working-men of this country, as their firm resolve to uphold themselves.

Still, unhappily, it is an undoubted fact that the most industrious men in large cities are too often unable to procure a fit dwelling, however able to pay for it and desirous to procure it. The houses have been built with no reference to such increasing wants. The idle and the diligent, the profligate and the prudent, the criminal and the honest, the diseased and the healthful, are therefore thrust into close neighbourhood. There is no escape. Is this terrible evil incapable of remedy? To discover that remedy, and apply it, is truly a national concern; for assuredly there is no capital of a country so worth preserving in the highest state of efficiency as the capital it possesses in an industrious population. There is a noble moral in a passage of Scott's romance, 'The Legend of Montrose.' A Highland chief had betted with a more luxurious English baronet whom he had visited, that he had better candlesticks at home than the six silver ones which the richer man had put upon his dinner-table. The Englishman went to the chief's castle in the hills, where the owner was miserable about the issue of his bragging bet. But his brother had a device which saved the honour of the clan. The attendant announced that the dinner was ready, and the candles lighted. Behind each chair for the guests stood a gigantic Highlander with his drawn sword in his right hand, and a blazing torch in his left, made of the bog-pine; and the brother exclaimed to the startled company—'Would you dare to compare to THEM in value the richest ore that ever was dug out of the mine?'

We may naturally pass from these considerations to a most important branch of the great subject of the expenditure of capital for public objects.

The people who live in small villages, or in scattered habitations in the country, have certainly not so many *direct* benefits from machinery as the inhabitants of towns. They have the articles at a cheap rate which machines produce, but there are not so many machines at work for them as for dense populations. From want of knowledge they may be unable to perceive the connexion between a cheap coat, or a cheap tool, and the machines which make them plentiful, and therefore cheap. But even they, when the saving of labour by a machine is a saving which immediately affects them, are not slow to acknowledge the benefits they derive from that best of economy. The Scriptures allude to the painful condition of the "hewers of wood" and the "drawers of water;" and certainly—in a state of society where there are no machines at all, or very rude machines—to cut down a tree and cleave it into logs, and to raise a bucket from a well, are very laborious occupations, the existence of which, to any extent, amongst a people, would mark them as remaining in a wretched condition. Immediately that the people have the simplest mechanical contrivance, such as the loaded lever, to raise water from a well, which is found represented in Egyptian sculpture, and also in our own Anglo-Saxon drawings, they are advancing to the condition of raising water by machinery. The oriental *shadoof* is a machine. In our own country, at the present day, there are not many houses, in situations where water is at hand, that have not the windlass, or, what is better, the pump, to raise this great necessary of life from the well. Some cottagers, however, have no such machines, and bitterly do they lament the want of them. We once met an old woman in a country district tottering under the weight of a bucket, which she was labouring to carry up a hill. We asked her how she and her family were off in the world. She replied, that she could do pretty well with them, for they could all work, if it were not for one thing—it was one person's labour to fetch water from the spring; but, said she, if we had a pump handy, we should not have much to complain of. This old woman very wisely had no love of labour for its own sake; she saw no advantage in the labour of one of her family being given for the attainment of a good which she knew might be attained by a very common invention. She wanted a machine to save that labour. Such a machine would have set at liberty a certain quantity of labour which was previously employed unprofitably; in other words, it would have left her or her children more time for more profitable work, and then the family earnings would have been increased.

Ancient Shadoof.

But there is another point of view in which this machine would have benefited the good woman and her family. Water is not only necessary to drink and to prepare food with, but it is necessary for cleanliness, and cleanliness is necessary for health. If there is a scarcity of water, or if it requires a great deal of labour to obtain it, (which comes to the same thing as a scarcity,) the uses of water for cleanliness will be wholly or in part neglected. If the neglect becomes a habit, which it is sure to do, disease, and that of the worst sort, cannot be prevented.

When men gather together in large bodies, and inhabit towns or cities, a plentiful supply of water is the first thing to which they direct their attention. If towns are built in situations where pure water cannot be readily obtained, the inhabitants, and especially the poorer sort, suffer even more misery than results from the want of bread or clothes. In some cities of Spain, for instance, where the people understand very little about machinery, water, at particular periods of the year, is as dear as wine; and the labouring classes are consequently in a most miserable condition. In London, on the contrary, water is so plentiful, that, as it appears from a return of the various water companies, the daily average of water-supply is sixty-two million gallons, being an average of about two hundred and two gallons to each house and other buildings, which amount to three hundred and ten thousand. This seems an enormous supply; but there are reasons for thinking that the quantity ought to be increased, and the arrangements made so perfect, that there should be a perpetual stream of water through the pipes of each house, like that through the arteries from the heart. The condition upon which the present water companies are allowed to continue their functions is, that they shall, before the expiration of another year, provide a larger and a purer supply. Yet, incomplete as these arrangements are, they are wondrous when compared with the water-supply of other times; and it is satisfactory to know that there are very few of our great towns which are not supplied as well as, and many much better than, London. There are very few large places in Great Britain where, by machinery, water is not only delivered to the kitchens and washhouses on the ground-floors, where it is most wanted, but is sent up to the very tops of the houses, to save even the comparatively little labour of fetching it from the bottom. The cost of this greatly varies in particular localities; and in most places the supply is afforded more cheaply than in the metropolis. There are natural difficulties in London, as in other vast cities, which have been chiefly created through the unexampled increase of the people. The sanitary arrangements of our great towns—the supply of water, the drainage—have followed the growth of the population and not preceded it. As the necessity has arisen for such a ministration to the absolute wants of a community, it has inevitably become a system of expedients. We are wiser now when we build upon new ground. We first construct our lines of street, with sewers, and water-pipes, and gas-pipes, and then we build our houses. What a different affair is it to manage these matters effectually when the houses have been previously built with very slight reference to such conditions of social existence!

As long ago as the year 1236, when a great want of water was felt in London, the little springs being blocked up and covered over by buildings, the ruling men of the city caused water to be brought from Tyburn, which was then a distant village, by means of pipes; and they laid a tax upon particular branches of trade to pay the expense of this great blessing to all. In succeeding times more pipes and conduits, that is, more machinery, were established for the same good purpose; and two centuries afterwards, King Henry the Sixth gave his aid to the same sort of works, in granting particular advantages in obtaining lead for making pipes. The reason for this aid to such works was, as the royal decree set forth, that they were "for the common utility and decency of all the city, and *for the universal advantage*," and a very true reason this was. As this great town more and more increased, more waterworks were found necessary; till at last, in the reign of James the First, which was nearly two hundred years after that of Henry the Sixth, a most ingenious and enterprising man, and a great benefactor to his country, Hugh Myddleton, undertook to bring a river of pure water above thirty-eight miles out of its natural course, for the supply of London. He persevered in this immense undertaking, in spite of every difficulty, till he at last accomplished that great good which he had proposed, of bringing wholesome water to every man's door. At the present time, the New River, which was the work of Hugh Myddleton, supplies more than seventeen millions of gallons of water every day; and though the original projector was ruined, by the undertaking, in consequence of the difficulty which he had in procuring proper support, such is now the general conviction of the advantage which he procured for his fellow-citizens, and so desirous are the people to possess that advantage, that a share in the New River Company, which was at first sold at one hundred pounds, is now worth three thousand pounds.

Before the people of London had water brought to their own doors, and even into their very houses, and into every room of their houses where it is desirable to bring it, they were obliged to send for this great article of life—first, to the few springs which were found in the city and its neighbourhood, and, secondly, to the conduits and fountains, which were imperfect mechanical contrivances for bringing it.

Conduit in Westcheap.

The service-pipes to each house are more perfect mechanical contrivances; but they could not have been rendered so perfect without engines, which force the water above the level of the source from which it is taken. When the inhabitants fetched their water from the springs and conduits there was a great deal of human labour employed; and as in every large community there are always people ready to perform labour for money, many persons obtained a living by carrying water. When the New River had been dug, and the pipes had been laid down, and the engines had been set up, it is perfectly clear that there would have been no further need for these water-carriers. When the people of London could obtain two hundred gallons of water for twopence, they would not employ a man to fetch a single bucket from the river or fountain at the same price. They would not, for the mere love of employing human labour directly, continue to buy an article very dear, which, by mechanical aid, they could buy very cheap. If they had resolved, from any mistaken notions about machinery, to continue to employ the water-carriers, they must have been contented with one gallon of water a day instead of two hundred gallons. Or if they had consumed a larger quantity, and continued to pay the price of bringing it to them by hand, they must have denied themselves other necessaries and comforts. They must have gone without a certain portion of food, or clothing, or fuel, which they are now enabled to obtain by the saving in the article of water. To have had for each house two hundred gallons of water, and, in having this two hundred gallons of water, to have had the cleanliness and health which result from its use, would have been utterly impossible. The supply of one gallon, instead of two hundred gallons to each house, would at present amount to 310,000 gallons daily; which at a penny a gallon would cost 1291*l.* per day; or 9037*l.* per week; or 469,724*l.*, or very nearly half a million, per year. Upon the assumption that one man, without any mechanical arrangement besides his can, could carry twenty gallons a day, thus earning ten shillings a week, this would employ no fewer than 18,074 persons—a very army of water-carriers. To supply ten gallons a day to each house would cost nearly five millions a year, and would employ 180,740 persons. To supply two hundred gallons a day would require 3,614,800 persons—a number exceeding the total population of London. The whole number of persons engaged in the waterworks' service of all Great Britain is under 1000.

Old water-carrier of London.

There is now, certainly, no labour to be performed by water-carriers. But suppose that five hundred years ago, when there were a small number of persons who gained their living by such drudgery, they had determined to prevent the bringing of water by pipes into London. Suppose also that they had succeeded; and that up to the present day we had no pipes or other mechanical aids for supplying the water. It is quite evident that if this misfortune had happened—if the welfare of the many had been retarded (for it never could have been finally stopped) by the ignorance of the few—London, as we have already shown, would not have had a twentieth part of its present population; and the population of every other town, depending as population does upon the increase of *profitable* labour, could never have gone forward. How then would the case have stood as to the amount of labour engaged in the supply of water? A few hundred, at the utmost a few thousand, carriers of water would have been employed throughout the kingdom; while the smelters and founders of iron of which water-pipes are made, the labourers who lay down these pipes, the founders of lead who make the service-pipes, and the plumbers who apply them; the carriers, whether by water or land, who are engaged in bringing them to the towns, the manufacturers of the engines which raise the water, the builders of the houses in which the engines stand—these, and many other labourers and mechanics who directly and indirectly contribute to the same public advantage, could never have been called into employment. To have continued to use the power of the water-carriers would have rendered the commodity two hundred times dearer than it is supplied by mechanical power. The present cheapness of production, by mechanical power, supplies employment to an infinitely greater number of persons than could have been required by a perseverance in the rude and wasteful system which belonged to former ages of ignorance and wretchedness.

When a severe frost chokes up the small water-pipes that conduct the useful stream into each house, what anxiety and trouble is there in every thoroughfare! The main pipes are not frozen; and the supply is to be got in pails and pitchers from a plug in the pavement, where a temporary cock is inserted. How gladly is this device resorted to! But imagine it to be the labour of every day, and what an amount of profitable time would be deducted from domestic employ!

Plug in a frost.

When society is more perfectly organized than it is at present, and when the great body of the people understand the value of co-operation for procuring advantages that individuals cannot attain, public baths will be established in every town, and in every district of a town. The great Roman people had public baths for all ranks; and remains of their baths still exist in this country. The great British people have only thought within these few years that public baths were a necessity. The establishment of public washhouses, in connexion with baths, having every advantage of machinery and economical arrangements, are real blessings to the few who now use them.

It is little more than thirty years since London was lighted with gas. Pall Mall was thus lighted in 1807, by a chartered company, to whose claims for support the majority of householders were utterly opposed. They had their old oil-lamps, which were thought absolute perfection. The main pipes which convey gas to the London houses are now fifteen hundred miles in length. There are, we believe, nearly a thousand proprietory gas-works in Great Britain. The noblest prospect in the world is London from Hampstead Heath on a bright winter's evening. The stars are shining in heaven, but there are thousands of earthly stars glittering in the city there spread before us: and as we look into any small space of that wondrous illumination, we can trace long lines of light losing themselves in the general splendor of the distance, and we can see dim shapes of mighty buildings afar off, showing their dark masses amidst the glowing atmosphere that hangs over the capital for miles, with the edges of flickering clouds gilded as if they were touched by the first sunlight. This is a spectacle that men look not upon, because it is common; and so we walk amidst the nightly splendours of the Strand, and forget what it was in the middle of the last century—the days of "darkness visible," under the combined efforts of the twinkling lamp, the watchman's lantern, and the vagabond's link.

The last, but in many respects one of the most useful of public works in Great Britain, to which a large amount of capital has been devoted, is the construction of sewers in our cities and towns. Popular intelligence and official power have been very slowly awakened to the performance of this duty. And yet the consequences of neglect have been felt for centuries. In 1290 the monks of White Friars and of Black Friars complained to the king that the exhalations from the Fleet River overcame the pleasant odour of the frankincense which burned on their altars, and occasioned the deaths of the brethren. This was the polluted stream that in time came to be known as Fleet Ditch, which Pope described as

> "The king of dykes, than whom no sluice of mud
> With deeper sable blots the silver flood."

London street-lights, 1760.

Fleet Ditch became such a nuisance that it was partly filled up by act of parliament soon after these lines were written. The Londoners had then their reservoirs of filth, called laystalls, in various parts near the river; and the pestilent accumulations spread disease all over the city. The system of sewers was begun in 1756, and from that time to the present several hundreds of miles of sewers have been constructed. But, alas, the Thames itself is now "the king of dykes," and the metropolis, healthy as it is, will never attain the sanitary state of which it is capable till the whole system of the outfall of the sewers is changed. The necessary work would involve the expenditure of millions. But the millions must be spent. In the mean time it is satisfactory to know that in towns of smaller population, where the evil is far less vast, and the natural difficulties of removal greatly less, the work of purification is going on rapidly. Public opinion has gone so strongly in the direction of a thorough reformation, that the duty can no longer be neglected. Every thousand pounds of public capital so expended is an addition to one of the best accumulations of national wealth.

www.ingramcontent.com/pod-product-compliance
Lightning Source LLC
Chambersburg PA
CBHW081617100526
44590CB00021B/3472